CABI Climate Change Series

Climate change is a major environmental challenge to the world today, with significant threats to ecosystems, food security, water resources and economic stability overall. In order to understand and research ways to alleviate the effects of climate change, scientists need access to information that not only provides an overview of and background to the field, but also keeps them up to date with the latest research findings.

This series addresses many topics relating to climate change, including strategies to develop sustainable systems that minimize impact on climate and/or mitigate the effects of human activity on climate change. Coverage will encompass all areas of environmental and agricultural sciences. Aimed at researchers, upper-level students and policy makers, titles in the series provide international coverage of topics related to climate change, including both a synthesis of facts and discussions of future research perspectives and possible solutions.

Titles Available

1. Climate Change and Crop Production
   Edited by Matthew P. Reynolds

2. Crop Stress Management and Climate Change
   Edited by José L. Araus and Gustavo A. Slafer

3. Temperature Adaptation in a Changing Climate: Nature at Risk
   Edited by Kenneth Storey and Karen Tanino

4. Plant Genetic Resources and Climate Change
   Edited by Michael Jackson, Brian Ford-Lloyd and Martin Parry

5. Climate Change Impact and Adaptation in Agricultural Systems
   Edited by Jürg Fuhrer and Peter J. Gregory

6. Livestock Production and Climate Change
   Edited by Pradeep K. Malik, Raghavendra Bhatta, Junichi Takahashi, Richard A. Kohn and Cadaba S. Prasad

7. Climate Change and Insect Pests
   Edited by Christer Björkman and Pekka Niemelä

8. Climate Change and Agricultural Water Management in Developing Countries
   Edited by Chu Thai Hoanh, Vladimir Smakhtin and Robyn Johnston

9. Climate Change Challenges and Adaptations at Farm-level: Case Studies from Asia and Africa
   Edited by Naveen Prakash Singh, Cynthia Bantilan, Kattarkandi Byjesh and Swamikannu Nedumaran

10. Climate Change Impacts on Urban Pests
Edited by Partho Dhang

11. Climate Change Impacts and Sustainability: Ecosystems of Tanzania
Edited by Pius Z. Yanda, Claude G. Mung'ong'o and Edmund B. Mabhuye

12. Climate, Ticks and Disease
Edited by Pat Nuttall

13. Climate Change, Adaptation and Gender
Mamta Mehar and Narayan Prasad

Climate Change, Adaptation and Gender

Policy, Practice and Methodological Underpinnings

Mamta Mehar

Senior Consultant, Athena Infonomics

and

Narayan Prasad

Professor (Economics), Indira Gandhi National Open University New Delhi, India

CABI

CABI is a trading name of CAB International

CABI
Nosworthy Way
Wallingford
Oxfordshire OX10 8DE
UK

Tel: +44 (0)1491 832111
Fax: +44 (0)1491 833508
E-mail: info@cabi.org
Website: www.cabi.org

CABI
WeWork
One Lincoln Street
24th Floor
Boston, MA 02111
USA

Tel: +1 (617)682-9015
E-mail: cabi-nao@cabi.org

A catalogue record for this book is available from the British Library, London, UK.

Library of Congress Cataloging-in-Publication Data

Names: Mehar, Mamta, author. | Prasad, Narayan (Scientist), author.
Title: Climate change, adaptation and gender : methodological underpinnings / Mamta Mehar, Narayan Prasad.
Description: Wallingford, Oxfordshire, UK ; Boston, MA, USA : CAB International, [2022] | Includes bibliographical references and index. | Summary: "This book offers a wide, in-depth study of the gender-climate change-agriculture nexus. Despite more than 40 years of research exploring this nexus, there is still ambiguity around the foundations, connections, and approaches for planning gender-inclusive climate policies. This volume aims to clear that ambiguity"-- Provided by publisher.
Identifiers: LCCN 2021060934 (print) | LCCN 2021060935 (ebook) | ISBN 9781789249897 (hardback) | ISBN 9781789249903 (ebook) | ISBN 9781789249910 (epub)
Subjects: LCSH: Women and the environment. | Climatic changes. | Environmental policy.
Classification: LCC GE195.9.M46 2022 (print) | LCC GE195.9 (ebook) | DDC 363.738746082--dc23/eng20220726
LC record available at https://lccn.loc.gov/2021060934
LC ebook record available at https://lccn.loc.gov/2021060935

ISBN: 978 1 78924 9897 (hardback)
978 1 78924 9903 (ePDF)
978 1 78924 9910 (ePub)

DOI: 10.1079/9781789249910.0000

Commissioning Editor: Ward Cooper
Editorial Assistant: Lauren Davies
Production Editor: Tim Kapp
Infographics: Pulkit Gandhi, Graphic Designer Associate, Athena Infonomics, India

Typeset by Straive, Pondicherry, India
Printed and bound in the USA by Integrated Books International, Dulles, Virginia

Contents

Foreword - Deepa Karthykeyan

Climate studies, conferences, and agreements are bringing to light how climate change may have a significant influence on our level of life, including food, clean drinking water, and sanitation. The Glasgow Climate Pact, 2021 (UNFCCC COP 26) draws attention to the fact that "human activities have produced roughly 1.1 °C of warming to date, that repercussions are already being felt in every region, and that carbon budgets commensurate with meeting the Paris Agreement temperature goal are currently modest and quickly depleting." Although the consequences of climate change may be seen in many sectors, agriculture is the most susceptible, threatening global food security and nutrition. However, in agriculture, women are more exposed and vulnerable to the effects of climate change because about two-thirds of the labour force in developing countries is female. According to WHO, climate shocks and natural disasters, in general, have a detrimental influence on gender equality because they increase women's vulnerability by reducing their access to resources, decision making power and mobility. Ingrained societal norms and socioeconomic institutions further accentuate gender inequities triggered by climate stresses and shocks, including exposing women and girls to gender-based violence such as sexual violence, human trafficking, child marriage, and other forms of abuse. Women and girls are also less able to seek relief and support in the aftershock, further jeopardising their livelihoods, welfare, and rehabilitation, and establishing a vicious cycle of susceptibility to future catastrophes.

A country's sensitivity to the impacts of climate change is essentially determined by how quickly and efficiently it can adapt to changes. As a result, adaptation is seen as the primary response option for mitigating the hazards posed by climate change. Adaptation methods that fail to account for the underlying gender inequality are likely to exacerbate social injustice and inequity, making adaptation less successful or even ineffective, and communities less resilient.

This is the primary emphasis of this book. The first author, through her extensive field experience, helps the reader unpack the specific ways in which climate change affects women and girls differently and ways in which we can equip them to adapt and cope with climate induced shocks and stresses through creative policies and programming. It will meet the needs of people interested in climate, gender, and agricultural research, as well as help policymakers working to develop climate smart policies, including gender sensitive policies. It will not just explore theories and methods to address climate concerns, but also include all elements related to gender, such as social, psychological, and practical aspects.

The book is a timely value addition to the scholarship and practice of gender intentional climate policy and programming as this year marks 30 years since the UNFCCC was first adopted and seven years since the Paris Agreement was agreed at COP21, where gender equality, empowerment of women and intergenerational equity was formally included in climate diplomacy for the first time.

Deepa Karthykeyan
Member of Technical Advisory Group, UN Global Partnership for Sustainable Development Data (GPSDD),
Member of Digital Water Programme Steering Committee, International Water Association (IWA)
Co-Founder Athena Infonomics, USA

Foreword - Rajiv Kumar

Climate change is a global emergency today. It needs all our policy attention and intensive community participation for us to succeed in reversing the environmental damage that has already been done. Climate change is already impacting all aspects of our lives including food, clean drinking water, and sanitation. Therefore, any book that focuses on global climate change is always welcome as it nudges us to think about this global problem.

Gender equality and tools to measure gender discrimination, the evolution of the concept of gender inclusiveness and its concerns, and the need to address the same by formulating gender-inclusive policymaking will all help to improve the extent of community participation in understanding climate change and our efforts to adapt to these changes and mitigate the impact as far as possible. Gender-based responses and responsiveness are crucial issues to understand the repercussions of climate change on the community. All these issues have been covered in this book in the right perspective.

This is an in-depth study of the gender-climate change-agriculture nexus. Gender equality and tools to measure gender discrimination, the evolution of the concept of gender inclusiveness and its concerns are all important to understand this nexus. This will contribute to the making of gender-inclusive policymaking. Despite more than 50 years since gender concerns were included in the discussion on exploring this nexus, there is still ambiguity around the foundations, connections, and approaches for planning gender-inclusive climate policies. The authors have made a commendable attempt to fill this gap.

This book will be useful to students, scholars and researchers working in the area of general studies, agriculture, climate change and rural development. It will also be a useful companion to practitioners, extension workers, and planners involved in designing new climate-resilient practices.

Rajiv Kumar
Chairman, Pahle India Foundation
Former Vice Chairman
NITI Aayog

Preface

"Climate change is Global. It would not exempt an atom from its impact and demands a sustainable development with increasingly limited resources amidst the looming challenges of the growing world."

The above statement is a contextual reflection of this book, which can be considered an evocative deliberation in understanding the concept of climate change and its impact on human resources considering cultural, regional, social, economic, financial, political, and geographical dimensions. Throughout this book, one can achieve insight into the various magnitudes of climate change's impact on various stages of gender needs.

In the last few years, farmers have been facing increasingly negative impacts of climate change and variability on their farm production, food security, water availability and livelihoods. The ways and scale in which climate impacts are experienced and responded to vary widely across countries, sectors, and cultural, environmental, economic, and social contexts. To minimize these negative impacts, numerous climate-resilience technologies and practices with the potential to withstand the changes already exist, promising yields which can reduce carbon emissions from agriculture. At this stage, it is important to understand the reasons for low adoption (and discontinuation in the following year) of climate-resilient technologies and practices. The degree to which people are affected by climate change impacts and to which they adapt mitigation strategies is partly a function of their social status, gender, poverty, power, and access to and control over resources. It would not be an exaggeration to say that the gender dimension of climate change research has not been a priority in the past few years. Despite growing recognition of the need for a gender-inclusive approach in climate-resilient approaches, for equity and productivity reasons, much agricultural and environmental research remains gender blind.

Recognising this gap, this book aims to provide pathways for research that adequately account for gender roles and relations as a foundation for better research, policy, and ultimately practice in climate resilient strategies. The book starts with an overview of climate change in relation to agriculture and its effect on food security. Next, it offers historical evaluation of international policy agreements for gender-inclusive climate policies with clarification of different buzz words used for 'climate' and 'gender'. The book reiterates that effective targeting of agriculture, food security and nutrition programs requires climate change adaptation and mitigation as a necessity. This further requires up-to-date and reliable information on the population's exposure, vulnerability, and need (whether individual (male/female of different age, class etc.), household, communities, or other administrative divisions such as regions, districts etc.). However, many nations or programs planned at the macro level do not collect and integrate such information at high temporal and spatial granularity levels in their vision. The book uniquely explores every dimension of climate change (i.e., people's beliefs or perceptions, knowledge, experience, coping, adaptation, and mitigation strategies) and links it with the gender dimension, including new theoretical and methodological approaches that go beyond the household as the unit of analysis and include approaches that include intersectional analysis. All these dimensions are envisaged in this book with considerable evidence-based facts at the global, national, and local levels.

About the Authors

Mamta Mehar

Mamta Mehar holds a PhD in Economics and has expertise in clientele-responsive agriculture and aquaculture innovative technologies, seed systems, digital agriculture, gender and climate change themes, interdisciplinary approaches, and innovative methods and tools to analyse and interpret data. During the past decade she has worked with different CGIAR (Consultative Group on International Agricultural Research) organizations and hence different food crops and programmes. Within these roles, and others, she has explored solutions to the inequalities embedded within gender roles and norms, which often result in the unequitable distribution of resources and hinder sustainable farm intensification. In 2017 she was awarded a Borlaug Fellowship by the US Department of Agriculture. She has also undertaken extensive Gender and Research Integrated Training (GRIT) organized by Pennsylvania State University (2017–18).

Narayan Prasad

Narayan Prasad has been Professor of Economics in the School of Social Sciences, Indira Gandhi National Open University, New Delhi, since 2006. He has also worked as the Director of the Research Unit and the Director of the Academic Coordination Division of the same. He has designed and developed research methodology courses for MPhil and PhD students (17 students have been awarded PhD degrees under his supervision), has published five books and has presented and published research papers in several international and national forums. He has delivered lectures in New York, Washington and Los Angeles, amongst others, and has been invited to deliver lectures on various aspects of research methodology for social sciences at various universities in India.

Acronyms and Abbreviations

CCA	Climate change adaptation
CCAFS	CGIAR (Consultative Group on International Agricultural Research) Research Program on Climate Change, Agriculture and Food Security
ccGAP	Climate change Gender Action Plan
CH_4	Methane
CGIAR	Consultative Group on International Agricultural Research
CO_2	Carbon dioxide
COP	Conference of the Parties
CRS	Climate-resilient strategies (strategies include actions and choices)
CSA	Climate-smart agriculture
FAO	Food and Agriculture Organization
FGD	Focus group discussions
FHH	Female-headed household
GAP	Gender Action Plan
GDP	Gross domestic product
GHG	Greenhouse gas
GESI	Gender equality and social inclusion
GIS	Geographic information systems
Ha	Hectare
ICT	Information and communication technology
IGWG	Inter-agency Gender Working Group
IPCC	Intergovernmental Panel on Climate Change
LLL	Laser land leveller
MHH	Male-headed household

M&E Monitoring and evaluation
MEAL Monitoring, evaluation, accountability and learning

NGO Non-governmental organization
N_2O Nitrous oxide
IPCC Intergovernmental Panel on Climate Change
PAR Participatory action research

PRA Participatory rural appraisal
REDD+ Reduced emissions from deforestation and degradation and forest conservation, plus the sustainable management of forests and the conservation and enhancement of forest carbon stocks
RMS Rice Monitoring Survey
SDG Sustainable Development Goal

UNFCCC United Nations Framework Convention on Climate Change
USAID United States Agency for International Development
WEDO Women's Environment & Development Organization
WHO World Health Organization
ZT Zero tillage

1 Introduction

Abstract

Over the years, climate change and climate shock have adversely affected the production of the agricultural sector (crops, livestock, forestry and fisheries) and damaged food security. However, in the context of the magnitude of effect and vulnerability, there are variations across locations, gender, age and status of food producers. This chapter clarifies the concepts related to 'climate' and 'gender'. The chapter expands and formalizes the conceptual foundations of the 'climate change–gender–agriculture' nexus, drawing on the theory and practices discussed in the remaining parts.

1.1 An Unprecedented Confluence of Pressures of Climate Change on Food Producers

Widespread, unprecedented and indisputable climate-induced events have threatened agricultural production, food security, water availability, biodiversity and energy security across the world. Climate-related risks as a result of either slow-onset or rapid-onset disasters have caused major losses of revenue to the agricultural sector with variation across crops and origin (region, country) of climate events (Box 1.1). Despite progress in recent decades, severe food insecurity (see Glossary) continues to affect 704 million people, globally, of which nearly 50% are in Asia followed by around 40% in Africa (FAO). The majority of the food producers in these regions are farmers with smallholdings and, ironically, are those who belong to the severe food-insecure section of society (Sibhatu and Qaim, 2017). However, in the context of the magnitude of effect and vulnerability, there are variations across locations due to the type and intensity of climate shocks and farmers' varying abilities to cope and adapt as cognition variables (such as social system, gender, age and economic status) influence behaviour, actions and gender.

DOI: 10.1079/9781789249910.0001

Box 1.1. Economic impacts of climate events.

Losses by disaster types:
Tropical storms: USD 9.65 million estimated total impact of the tsunami in 2014 on the agricultural sector of Thailand (ADPC, 2006).
Drought: In 2015, California drought cost the agricultural sector USD 2.7 billion (CDP, 2015). The Horn of Africa drought in 2011 cost the economy of Kenya up to 1% of its gross domestic product (GDP) (Demombynes and Kiringai, 2011).
Floods: Agricultural impacts of floods caused national economic losses that ranged from 3.5% to 6.5% of GDP in Malawi (Pauw *et al.*, 2012).

Crop-wise losses between 2005 and 2016 due to natural disasters compiled by the FAO (2018):
- Over the past ten years, the production of roots and tubers – such as potatoes, sweet potatoes, cassava and yams – sustained the highest loss in Africa, amounting to just over USD 9 billion.
- Cereal and livestock production loss followed closely at USD 5 billion and USD 4 billion, respectively.
- Disasters in Asia had a serious impact on fruit and nut production (loss of USD 7.3 billion), livestock production (loss of just over USD 6 billion) and vegetable production (loss of about USD 5 billion).
- For Latin America and the Caribbean, the production of leguminous crops such as beans, lentils and chickpeas was mostly affected, causing a loss of just under USD 8 billion.

According to the Intergovernmental Panel on Climate Change (IPCC), 'climate change refers to any change in climate over time, due to natural variability or as a result of human activity' (IPCC, 2007a, p. 6). At the global level, the IPCC and several other studies have projected the expected decrease in yields. Climate change has a two-way cause-and-effect relationship with agriculture. Climate change affects agriculture in several ways (Fig. 1.1). For example, the climate change effect on livestock is in terms of a reduction in the quantity and quality of feed (Rojas-Downing *et al.*, 2017), a reduction in timber and non-timber forest products (Wunder *et al.*, 2018), and a subsequent fall in both the quantity and distribution of fish and other marine resources (Cheung *et al.*, 2016). Except for agriculture, there is limited literature available for other sectors.

Scientific evidence for climate change projection suggests that soon there will be a significant reduction in agricultural productivity, especially in countries at lower latitudes (Stevanović *et al.*, 2016).

Forestry and land use comprises: forest conversion (38%), peat degradation (11%), biomass fires (1%). Livestock, especially cattle, produce methane (CH_4) as part of their digestion. This process is called 'enteric fermentation'. The other non-CO_2 greenhouse gas (GHG) is nitrous oxide (N_2O).

The impact of, consequences for and solutions to the climate change and agriculture relationship are explored in numerous studies. In Africa and South Asia, major grains including wheat, maize and sorghum are projected to suffer a mean yield reduction of 8% by 2050, with significant reductions projected for wheat

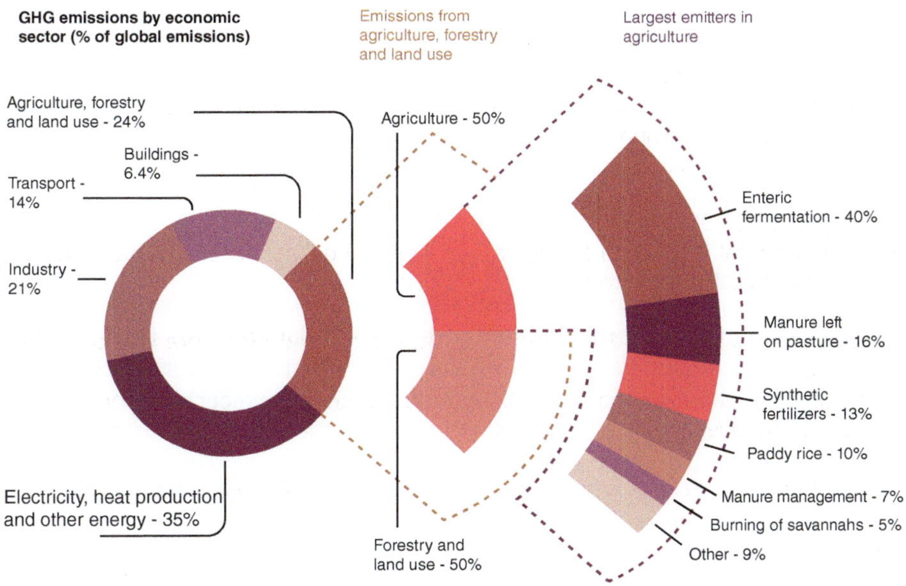

Fig. 1.1. Agriculture and land use contribution to global GHG emissions. (CDP, 2015)

(−12%), maize (−7%), sorghum (−3%) and millet (−9%) (Knox *et al.*, 2012). Yohannes (2016) has provided an assessment of the reviewed articles on the relationship between climate change and agriculture. The key inferences drawn from his paper are as follows:

- The impact of climate change tends to be more severe for countries with higher initial temperatures, areas with marginal or already degraded lands, and lower levels of development with little adaptation capacity.
- Developing countries' dependence on agriculture makes their situation more critical. Additionally, the contribution to GHG is primarily because of inadequate manure management, improper use of agrochemicals and mismanagement of land.
- Sustainable and organic agricultural systems can help reduce agricultural GHG emissions through energy conservation, lower levels of carbon-based inputs, lower use of synthetic fertilizer, and other features that minimize GHG emissions and sequester carbon in the soil.

Although climate change has global impacts, the nature and magnitude of the impacts are not homogeneous; rather, they differ across different segments of society. It may have profoundly different impacts on households of varying capabilities even in the same climate-affected area. Marginalized groups such as women, children, the elderly and the impoverished have less access to and control over resources and therefore are more severely impacted by climate change (Dankelman, 2002). Climate-related disasters often intensify existing social and gender inequalities and economic poverty (Brody *et al.*, 2008). Women and men perceive and experience climate change in different ways because of their distinct socially constructed gender roles, responsibilities, status and identities, which result in varied coping strategies and responses (Lambrou and Nelson, 2010; FAO, 2011). How households perceive a climate shock and react to it is, *ceteris paribus*, found to be significantly affected by the gender of the action taker and decision maker within a household (Mehar *et al.*, 2016). Since men and women play different roles, they often face very different cultural, institutional, physical and economic constraints, many of which are rooted in systemic biases and discrimination (ILO, 1998). Thus, women and men perceive and assess climate change differently (Table 1.1). The issues in Table 1.1 draw on Delaney and Sharder (2000, p. 14). To understand the role of women in adaptation to climate change, we need to understand power relations between women and men and the way climate change can exacerbate and widen these relations (Brody *et al.*, 2008).

The literature on gender and climate change echoes the importance of including gender considerations in climate-resilience strategies. Indeed, several agreements (at international, national and local level) have been created with a gender-responsive approach for climate mitigation. The most recent are the establishment, adoption and extension of the Lima Work Programme of the Gender Action Plan (GAP).

Table 1.1. Gender-differentiated responses and experiences in climate events. (From: the author)

Issues	Men	Women
Social stereotype outlook	Masculinity, power, household responsibility (as breadwinner), men seen as mobile	Weak, household responsibility (as caretaker and custodian of food and water), limited exposure and barrier for socialization
Conflict	Conflict over resources in society	Conflict over resources in household
Vulnerability	Anxiety, distress and suicide Displacement/temporary migration, forced to live alone, face urban poverty and informal settlements	Mortality, lack of survival skills (like swimming, climbing trees), forced to stay home, greater female headship, difficult to find employment opportunities post-disaster, reduced education opportunities as compared to men
Needs and priorities	In search-and-rescue, protector of family and community	Food and shelter
Coping strategies	Temporary migration, off-farm activities, borrowings	Selling household goods including jewellery, eating less
Programmes	Technology – male bias	Women's programmes/men exclusion Women not properly presented in policies

1.2 Demystifying Gender and Climate Change Lingua

In the context of the objective of examining the perception of gender roles in times of climate events, it is important to first understand the meaning of the concepts and voluminous terminology used by researchers and practitioners (Fig. 1.2).[1] The concepts in this section are mainly in the context of gender and climate change. The other concepts are presented in the Glossary or the relevant chapters in which they need to be explained.

1.2.1 Climate-related lingua

With the emergence of climatic events and people's awareness of them, the language or 'lingua' used to describe these is also evolving. The often-used terms are: climate variability, climate change, weather, global temperature, global warming, climate extreme, carbon neutral, carbon footprint, environmental justice, climate hazard, climate crisis, climate shock, climate risk, climate change adaptation, climate change mitigation, disaster, and disaster risk reduction, amongst others. Often the terms are used interchangeably and can be confusing. We have tried to clarify them in this section.

Weather refers to local conditions on the scale of minutes, hours, days and even months to years: for example, late arrival of winter, warmer winter as compared to another year, etc.

Climate change is a long-term shift in global or regional weather conditions as compared to the average weather conditions, or, more rigorously, as the statistical description in terms of the mean and variability of relevant quantities over a period ranging from months to thousands or millions of years. The IPCC (2007b, p. 78) defined climate change as a 'change in the state of the climate that can be identified by changes in the mean and/or variability of its properties that persists for an extended period, typically decades or longer'. The UN defined climate change as 'a change of climate which is attributed directly or indirectly to human activity that alters the composition of the global atmosphere and which is in addition

Fig. 1.2. Word cloud for climate change and gender lingua. (From: the author)

to natural climate variability observed over comparable periods' (United Nations, 1992). They also mention that it affects food quality and availability.

The difference between weather and climate change is a matter of time. Climate variability and climate change difference are also linked with the timescale of observed variations.

Climate variation or *variability* is a natural phenomenon that is dynamic and varies at a global scale of time and space (Ribot *et al.*, 1996). Climate variability, beyond individual weather events, refers to variations in the mean state and other statistics (such as standard deviations, statistics of extremes) of the climate on all temporal and spatial scales beyond that of individual weather events (IPCC, 2012, p. 257). This term will be used quite often in the book to reflect the understanding of the following parameters:

1. Temperature: this includes higher (lower) temperatures, heatwaves, hot months are getting hotter, cold months are getting colder, the average temperature is decreasing (increasing);
2. Precipitation: more (less) rainfall, higher intensity of rainfall, earlier (later) start (stop) of rainfall, untimely rainfall;
3. Extreme events: prolonged (frequent) droughts and/or floods;
4. Seasonal pattern: shift in timings of seasons, uneven weather fluctuations, increase (decrease) in the winters.

Climate extreme is defined as the occurrence of a weather or climate variable above (or below) a threshold value near the upper (or lower) ends of the range of observed values of the variable (IPCC, 2012, p. 257). Natural disaster events such as drought, floods and cyclones are referred to as 'climate extreme'. Other variables such as erratic rainfall, heatwaves, strong winds, hailstorms and others beyond a threshold (in terms of intensity) are also climate extreme.

Global warming refers to the overall increase in worldwide temperatures caused by rising greenhouse gas emissions. Climate change is different to this, as climate change encompasses the many long-term changes, globally, that people are experiencing.

Carbon footprint is the amount of carbon dioxide emitted by an individual or organization in a given period, or the amount of carbon dioxide emitted during a product manufacturing process.

Climate shock and *Climate crisis* are punchier and emotional terms to describe climate change, climate extreme and their consequences. In the literature, they are often referred to as any unexpected experienced past events that have caused a big problem or significant (negative) impact on farmers' production, livelihood and lives. In the literature, animal disease, serious pest damage to crops and market shocks also represent climate shocks.

Climate risk refers to 'risk assessments based on formal analysis of the consequences, likelihood and responses to the impacts of climate change and how societal constraints shape adaptation options' (Adger *et al.*, 2018).

Risk involves uncertainty, often referring to situations where the outcome (effects/implications) of an activity or situation is unknown and often having negative or undesirable consequences.

Climate-related risks include 'those from potential acute or chronic natural disasters, change in climate patterns and the related technology, market, legal and changes in government policy risks'.[2]

Climate change adaptation, mitigation, and climate-smart agriculture concepts are defined and discussed in Part IV.

Environmental justice calls for the fair treatment and meaningful involvement of all people regardless of race, colour, nationality or income concerning the development, implementation and enforcement of environmental laws, regulations and policies, defined by USEPA (United States Environmental Protection Agency). Recognizing the need, at an international level, several agreements and commitments are made and revised from time to time. A snapshot of these can be viewed in Box 1.2.

1.2.2 Gender-related lingua

Analogous to 'climate', there are many linguae used around 'gender'. We focus on terms quite often used in climate change–agriculture–gender linkages. The terms are gender, gender identities, sex, gender-equal, gender equality, gender justice, gender-responsive, gender inclusion, gender needs, gender roles, gender mainstreaming, gender planning, gender analysis, gender discrimination, gender-responsive budgeting, gender-disaggregated data, gender diversity, gender stereotypes, gender blindness, gender bias, gender-transformative, gender-accommodating, gender perception, and others. These linguae come from the intensity of words used in gender and climate literature and in climate programme-planning by scholars.

'Gender' is often used interchangeably with 'sex', which generally refers to the biological differences between female and male. However, gender refers to the socially constructed[3] rather than biologically determined roles and relations between women and men, which include different responsibilities of women and men in a given society or culture at a specific location and time (Okin, 1989; Esplen, 2009; IEWM, n.d.). As stated by FAO, gender refers not to male and female but to masculine and feminine – that is, to qualities or characteristics that society ascribes to each sex (FAO, n.d.). People are born female or male but learn to be women and men in their respective culture and society. As gender understanding lies in the roots of the socially constructed norms, roles and relations, its definition is highly context-specific; it can vary deeply both within and across cultures. Bound to social norms are gender-stereotypical prejudices about the division of labour, how men and women should be treated, and gender discrimination (refers to an exclusion or restriction that prevents a person enjoying their full human rights) in access to and control over resources, in decision-making power, etc.), which leads to gender bias and gender exclusion from many programmes and policies. The resources referred to here are:

- economic resources: land or equipment;
- political resources: representation, leadership, legal structures;
- social: childcare, family planning, education, market barriers due to mobility constraint on women; and
- time: a scarce resource for women.

Box 1.2. Climate change timeline of actions.

1957: First warning issued by US oceanographer Roger Revelle who discovered that seawater will not absorb all the additional CO_2 entering the atmosphere.

1972: First UN environment conference in Stockholm where climate change registers attention; as a result, UNEP (United Nations Environment Programme) formed.

1975: US scientist Wallace Broecker uses the term 'global warming' in a scientific paper titled 'Climatic change: Are we on the brink of a pronounced global warming'.

1979: The first World Climate Conference (WCC) takes place in Geneva.

1987: Montreal Protocol – agreement made to restrict the production and consumption of substances depleting the ozone layer.

1988: IPCC established to provide policymakers with regular scientific assessments on the current state of knowledge about climate change.

1990: IPCC's first assessment report (main finding was temperature had risen by 0.3–0.6°C over the last century, the release of which shaped the agenda for the second WCC.

1991: First meeting of the Intergovernmental Negotiating Committee (INC) takes place.

1992: At the Earth Summit in Rio, the United Nations Framework Convention on Climate Change (UNFCCC) is opened for signature along with its sister Rio Convention, the UN Convention on Biological Diversity and the UN Convention to Combat Desertification

1994: UNFCCC enters into force.

1995: First Conference of the Parties (COP1) takes place in Berlin.

1997: Kyoto Protocol formally adopted in December at COP3. The Protocol legally binds developed countries to emission-reduction targets by an average of 5% by the period 2008–2012.

2001: Release of IPCC's Third Assessment Report. Bonn Agreements adopted, based on the Buenos Aires Plan of Action of 1998. Marrakesh Accords adopted at COP7, detailing rules for implementation of the Kyoto Protocol, setting up new funding and planning instruments for adaptation, and establishing a technology transfer framework. The Special Climate Change Fund (SCCF) was established to 'finance projects relating to: adaptation; technology transfer and capacity building; energy transport, industry, agriculture, forestry and waste management; and economic diversification'.

2005: Entry into force of the Kyoto Protocol. The first meeting of the Parties to the Kyoto Protocol (MOP1) takes place in Montreal, under Kyoto Protocol requirements.

2009: Copenhagen Accord drafted at COP15 in Copenhagen. Countries later submitted emissions reduction pledges or mitigation action pledges, all non-binding.

2007: IPCC's Fourth Assessment Report released. Climate science entered into popular consciousness.

2010: Cancun Agreements drafted and largely accepted by the COP at COP16. Through the Agreements, countries made their emission-reduction pledges official in what was the largest collective effort the world has ever seen to reduce emissions in a mutually accountable way. Green Climate Fund established.

2011: Durban Platform for Enhanced Action drafted and accepted by the COP at COP17. In Durban, governments clearly recognized the need to draw up the blueprint for a fresh universal, legal agreement to deal with climate change beyond 2020, where all will play their part to the best of their ability and all will be able to reap the benefits of success together.

2012: Doha Amendment to the Kyoto Protocol is adopted by the CMP at CMP8.

2013: Green Climate Fund and Long-Term Finance, the Warsaw Framework for REDD+ and the Warsaw International Mechanism for Loss and Damage.

Continued

Box 1.2. Continued.

2015: Sendai Framework for Disaster Risk Reduction set clear targets and priorities for action to prevent new, and reduce existing, disaster risks. Countries submit their first round of climate pledges (INDCs) in the Paris Agreement (COP21).
2017: Concrete Climate Action commitments to achieve the 2030 Agenda Sustainable Goals.
2018: Agreement to bring the Paris climate deal to life; parties agreed to convene a 'facilitative dialogue' to 'take stock' of collective efforts of countries in relation to long-term agreement and to inform the preparation of the next round of pledges.
2019: Release of *Global Landscape of Climate Finance* report, which is a comprehensive overview of global climate-related primary investment.
2021: Virtual Climate Adaptation Summit hosted by the Netherlands. UNFCCC (COP26) in Glasgow.

The constraints in accessing productive resources, markets and services are usually more severe among women farmers, farmworkers and rural entrepreneurs than among men. These constraints lead to discrepancies or gaps in opportunities, status and rights, which can be defined as a 'gender gap'. This gender gap often affects women's productivity and reduces their contribution to the output of the agricultural sector (FAO, 2011). Gender gaps lead to a lack of gender balance, gender equity and gender equality in policies. Gender equality is based on the idea of opportunities. Gender equality is when men and women of all ages enjoy equal rights, opportunities, resources and rewards. Equal treatment does not mean the *same* treatment. Gender equity recognizes that women and men have different needs, preferences and interests and may require different treatment. Gender equity and equality must be pursued in a complementary manner, even though gender equality is the ultimate goal. The development challenge in every case is to identify barriers to opportunities that exist and custom-design the adjusted interventions that will lead to equality of outcomes.

The IPCC considers equity to be:

> The principle of being fair and impartial, and a basis for understanding how the impacts and responses to climate change, including costs and benefits, are distributed in and by society in more or less equal ways. It is often aligned with ideas of equality, fairness and justice and applied with respect to equity in the responsibility for, and distribution of, climate impacts and policies across society, generations, and gender, and in the sense of who participates and controls the processes of decision making.
>
> (IPCC, 2019)

At times, special treatment, affirmative action and positive discrimination are required to compensate for the historical and social disadvantages of women, such as empowering women by special provisions in jobs and education so that we can achieve gender equity.

'Empowerment is about the process by which those who have been denied the ability to make strategic life choices acquire such an ability' (Kabeer, 1999, p. 435).

A required step, 'gender planning', refers to designing gender-responsive planning processes, taking into account the different roles, needs and priorities of women and men in the formulation of strategies and plans. However, most of the time, they are gender-blind or gender-neutral. Gender-blind ignores the issues and needs of men and women and does not include the gender dimension in projects or programmes. Gender-neutral is where a project or programme is not concerned with human activities and does not affect people; for example, forecasting the weather or monitoring changes in glacier melting.

To achieve gender balance, gender planning with gender-specific, sensitive, responsive and, in some cases, transformative action is required. These are discussed in detail in Chapter 3.

> The process of assessing the implications for men and women of any planned action, including legislation, policies, and programs, in any area and at all levels is defined as gender-mainstreaming. It is a strategy for making the concerns and experiences of women and men an integral part of the design, implementation, monitoring, and evaluation of policies and programs in all political, economic, and societal spheres, so that they benefit equally and inequality is not perpetuated.
>
> (ECOSOC, 1997, p. 28)

To understand current social policy, environment gender analysis or assessment should be done to examine the extent of gender-planning responsiveness to strategic and practical gender needs.

Gender analysis is a systematic way of collecting and analysing sex-disaggregated information for predefined indicators and themes (Reeves and Baden, 2000). This includes, but is not limited to, developing strategies for gender and social inclusion and avoiding marginalization. It needs to be integrated throughout the whole design cycle, from proposal and data collection to analysis, write-up, and monitoring and evaluation (Mehar and McDougall, 2018). Gender analysis has the capacity to distinguish both the differences and commonalities in social perceptions with regard to the two biological sexes. To collect adequate sex-disaggregated data, Doss and Kieran (2014) have introduced guidelines which provide clarification regarding gender and sex, unit of analysis, who to interview, asking the 'who' question, and context. The reader planning to conduct gender-analysis research is strongly encouraged to read this guidelines document: 'Gender analysis is a systematic analytical process for organizing, collecting, analysing and interpreting qualitative and quantitative information that examines gender relations in a particular context, ranging from households to communities to nations' (Mehar and McDougall, 2018).

Practical gender needs (PGN) are the immediate needs that address the practical interest of men or women within their socially accepted roles, often in response to the immediate perceived necessity to assist their survival (e.g. water, shelter, food). Strategic gender needs (SGN), on the other hand, are those related to gendered divisions of labour distribution, power and control, which aim to challenge existing gender roles as a means to greater equality. They might include issues such as legal rights, domestic violence, equal wages, etc. Due to the political nature of SGN, PGN is the preferred entry point for policy implementors, including NGOs and women's organizations.

To evaluate the level of planning and integrating gender, the CGIAR Program of Work and Budget guided CGIAR science leaders and gender researchers to score their programmes for budgeting and reporting in two major categories: integrated (applied) gender research and strategic gender research (CGIAR-IEA, 2017). Integrated (applied) gender research, 'integrates consideration of gender into technical research which is the principal topic of study, for example plant breeding, aquaculture, postharvest technology development, systems intensification' (CGIAR, 2015). Strategic gender research 'studies gender as the primary topic in a social analysis designed to understand what the implications of gender are for agriculture' (ibid.).

The terms in this sub-section are introduced along with the major programmes during the evolution of gender (Box 1.3). The timeline of gender engagement in climate policies is presented in Chapter 3. Efforts and initiatives mentioned in Box 1.3 turn to women's economic empowerment (WEE) and have led to the development of multiple empowerments and gender inequality indexes including: the Gender Development Index and the Gender Empowerment Measure developed by the United Nations Development Programme (UNDP, 1995), the Global Gender Gap Index from the World Economic Forum (Lopez-Claros and Zahidi, 2005), the Gender, Institutions and Development Database (GID-DB) (Jutting *et al.*, 2008) and the Women's Empowerment in Agriculture Index from USAID, IFPRI and the Oxford Policy and Human Development Initiative (Alkire *et al.*, 2013).

1.3 The Book's Organization

1.3.1 Aim of the book

There is no denying that curbing GHG emissions (or achieving net-zero agriculture emissions) is essential in order to keep global temperature rise below 1.5°C. Co-ordinating the urgent need for transformation is unquestionable in order to create climate-resilient strategies (CRS) to reduce GHG emissions as well as adverse climate impact on farmers' lives. However, the attitude, scale and magnitude of adoption of CRS varies due to diversity of risk, resilience and levels of development of individuals, households, communities and nations. Additionally, social stressors (such as age, gender, class, caste, region, wealth status, etc.) escalate the differences in the adoption of CRS. Gender inequality is seen more starkly against the onslaught of climate event impacts. This leads to increasing efforts to consider gender mainstreaming in climate-change policies, actions and research, especially after the Paris Agreement 2015. 2020 marked the 25th anniversary of the first momentum (i.e. Beijing Declaration and Platform for Action) of attention towards including gender concerns. All these efforts recognized the relevance of gendered climate change research and this book aims to bring gender concerns to the fore. It highlights the need for a synergetic and transformative process towards climate adaptation and mitigation strategies with gender-inclusive approaches. The book provides information globally. It is the first to uncover all aspects of gender and climate change using theoretical and practical models. It is constructed around three themes:

Box 1.3. Progression of gender rights.

1911: First celebration of International Women's Day.
1915: On 28 April, the first International Congress of Women held, later known as Women's International League for Peace and Freedom.
1960s: Considered the era of Women in Development (WID), created with limited dimension to human rights.
1970: The era of women, environment and development; assumption: women have an innate connection to nature (ecofeminism) so should be targeted.
1970s: Boserup (1970) investigated the role of women in the economic and social transformation of poor countries.
1975: First World Conference on Women held in Mexico.
1976–85: Declared the decade for women by the UN.
1979: UN adopted the Convention on the Elimination of All Forms of Discrimination against Women (CEDAW), also known as International Bill of Rights for Women.
1980s: Considered the era of Gender and Development (GAD) when several development agencies like the World Bank added gender to their development agendas.
1982: First gender-centric CGIAR (Consultative Group on International Agricultural Research) programme started. Women and Rice research programme started by Janice Jiggins.
1990: Longwe Women's Empowerment Framework developed.
1993: Moser developed a Gender Planning Framework.
1993: World Conference on Human Rights, Vienna. The declarations urged governments and organizations to use gender-specific data in their deliberations and findings related to human rights.
1995: Gender mainstreaming officially agreed and adopted at Fourth World Conference of Women.
1995: Beijing Platform for Action stressed gender mainstreaming as key to women's empowerment.
1995: UN World Conference discussion around feminization of agriculture and poverty led by Agnes Quisumbing (IFPRI).
2000: Millennium Development Goals (MDGs) include gender equity as key objective.
2006: Global Gender Gap Index is a framework for capturing the magnitude and scope of gender-based disparities and tracking their progress. It focuses on three basic concepts: measuring gaps in outcome variables (such as health, education, economic participation and political empowerment); gaps in input variables (such as specific policies, culture or customs); and gender equality and women's empowerment (Hausmann, 2006).
2006: Global Gender Gap Index countries should focus on gender equality rather than women's empowerment (Hausmann *et al.*, 2006).
2007: At Bali conference, the voice for gender climate justice is raised.
2009: *Gender in Agriculture: A Sourcebook* was published by the World Bank, FAO and IFAD, which included numerous gender issues of agricultural development.
2012: FAO produces statistics of women in agriculture – *State of Food and Agriculture: Women in Agriculture*.
2014: IPCC's Gender in Climate Change – a set of guidelines on sex-disaggregated data collection developed.
2015: Sustainable Development Goal (SDG) 5: Achieve gender equality and empower all women and girls.
2017: Women's March on Washington, 21 January – the largest international mass demonstration in support of women's rights.

Continued

Box 1.3. Continued.

2018: Efforts to bring gender philanthropy to the main stage; large funders commit to fund gender equality globally.
2020: 25th anniversary of Fourth World Conference on Women and the adoption of Beijing Declaration and Platform for Action.
2021: Gender Equality Forum launched a five-year action plan to achieve irreversible progress towards gender equality.

- gender and climate issues which need to be addressed as early as possible;
- identifying what issues are researchable; and
- identifying and critically evaluating appropriate methodology to address these issues.

The prime objectives of the book are to uniquely explore every dimension of climate change (i.e. people's beliefs or perceptions, knowledge, experience, coping, adaptation and mitigation strategies) and link it with gender, including new theoretical and methodological approaches that go beyond the household as the unit of analysis and include approaches that include intersectional analysis. Each of the chapters draws on theory and concepts related to agricultural, rural or sustainable development. Several of the chapters also engage with new ways of analysing gender and climate change in an interdisciplinary way. The book includes procedural details of social research together with lessons learned and recommendations from the people involved in the study. The literature review brings critical arguments for gender inclusion, comparisons and commonalities across studies with respect to adoption of climate-resilient practices, especially by smallholder farmers. Lastly, it provides methodological prescriptions for ongoing and future research. We would like to echo that the students will learn not only about major themes of research but also different methodologies, ranging from review methods to mathematical models, conceptual frameworks and empirical analysis.

All in all, the contents of the study are presented in six parts that can be summarized as: background and vision of the book (Part I: Setting the Context); narrating gender and climate change theory and policies (Part II: Gendering Climate Change); exploring people's understanding of climate change (Part III: A Gender Analysis in Climate Events); reflections from the field (Part IV: Gender Responses to Climate Changes); from the field back to theory (Part V: Gender and Social Inclusion Dimension in Farming System Analysis); takeaways and good practices (Part VI: Methodological Underpinnings and Prescriptions for Future Research). Relevant examples, wherever possible, from global to local research are provided.

Part I: Setting the Context. This part includes an introductory chapter that establishes the conceptual and theoretical understandings that guide the need for this book and set a framework for discussion in the subsequent parts.

Chapter 1 presents an overview of the climate change and climate shock effect on the agricultural sector (crop, livestock, forestry, capture fisheries and aquaculture) and hence the effect on food security. It clarifies the different lingua available and used for 'climate' and 'gender' and summarizes the evolution of both climate and gender within the context of international policy agreements and debates. Lastly, the chapter synthesizes the framing of the book and discusses its contribution, formulation (from idea to print), targeted audience and how it should be used.

Part II: Gendering Climate Change. Over recent decades, 'gender issues' in general and their connection with climate change have been explored widely. The issues are not only researched, but also several organizations have adopted a 'gendered approach' in terms of gender-integrating technology and training. Despite recent advances in terms of climate and gender research, until recently there has been limited evidence or experimental studies explaining gender in climate change. This is partly due to studies having looked for different dimensions with different methodologies. This part aims to provide a theoretical and empirical synthesis of the connections between gender and climate change.

Chapter 2 draws linkages between gender and climate change in the agricultural sector using interpretative paradigms.

Chapter 3 explores the evolution of gender inclusiveness in climate policies in major international agreements, and documents the institutional mechanisms and structural barriers in policy implementation. This chapter also identifies the misconceptions and concerns for gender inclusiveness.

Part III: A Gender Analysis in Climate Events. Gender and climate change are not directly linked, though their interaction is complex and dynamic. The complexity can be understood by unpacking the gendered impact of climate change, perception of climate change and different ways to believe and adapt for reducing the impact. Addressing climate change from a gender perspective, therefore, is concerned with engaging both men and women to ensure that their perceptions, interests, needs and priorities are given equal weight in planning and decision making. This part aims to shows how the psychological dimensions as to why adaptation of agriculture to climate change requires to be viewed through the gender lens and, at a broader level, micro-studies can be linked to other available data sets using the triangulation approach.

Chapter 4 builds on the narrative literature review which compiles the issues that point to the crucial role of gender when understanding the perception, knowledge and experience of climate change and its impact.

Chapter 5 provides descriptive statistics relevant to the psychological dimensions of the previous chapter. Here secondary data sources are used and a triangulation approach is adopted with the aim to support the use of different data sets for better understanding and external validity.

Part IV: Gender Responses to Climate Changes. This part works towards exploring gender-differentiated responses to climate change in terms of coping, adaptation or mitigation strategies, how farmers perceive environmental risk and how they plan to cope with it. There is a need to examine more micro-level studies aimed at the gender dynamics of stress. Have cultural barriers broken in the face of climate shocks (drought or flood)? Have they (cultural barriers) changed in other minor shocks (pest or disease attack)? Are they transferring as climate is promising less yield and male migration (seasonal or temporal) for non-agricultural work and females are left behind for field activities and the family? The gender parameters pave the way to differential understanding, impact on male and female farmers, as well as approaches followed by them to escape from the risks incurred.

Because many climate projects focus on one or more development sectors and many strategies can help to reduce climate risk, we have summarized the strategies available in Chapter 6. The chapter also discusses trade-offs and synergies among major strategies (such as coping, adaptation, mitigation and climate-smart technologies).

Chapter 7 describes the conceptual understanding of 'climate smart agriculture'. Evidence from South Asia and Africa are used to understand the gender dimension in the context of climate-smart agricultural practices and technologies.

Chapter 8 provides evidence for coping strategies highlighting preferences and choices for men and women using a conceptual literature review. The chapter also provides concise illustrative case studies for the different themes emerging during a conceptual literature review. It concludes by providing different variables which influence gender choices, and directs how and why these could be incorporated into design and implementation of gender best practice.

Part V: Gender and Social Inclusion Dimension in Farming Systems Analysis. The fact that climate has a different impact on men and women and they both have different priorities and strategies to cope with it has been well recognized. Part of the reason for this difference is due to differences in roles and responsibilities in agriculture. The other reason, often overlooked, is the participation of men and women in different activities across different stages of farming. This is crucial to understand for two reasons: first, climate-resilient strategies are relevant to different crop stages; and, second, which crop-stage climate shock occurs is uncertain. Chapter 9 explores this argument in detail.

The majority of the literature and evidence pertains to crops, though these remain anecdotal or based on case studies or some regions, with less evidence from cross-country or global inferences. However, there exist cross-sectoral synergies between and among crops, livestock, forestry, capture fisheries and aquaculture, with the potential of reducing climate risk by integrated activities or strategies. Chapter 10 discusses this aspect in detail. It also explores climate-smart practices with an entrepreneur mindset and discusses the effect of climate change and climate-resilient practices on health and nutrition of rural people.

Part VI: Methodological Underpinnings and Prescriptions for Future Research. This part builds on the information and gaps in gender and climate research explored in all of the previous parts. It aims to discuss gender concept, methodology and methods and approaches to gender research from a practical perspective. This, in consequence, considers ideal versus real-world contingencies in gender research, covering the issues and dilemmas that the gender researcher is likely to encounter. The goal is to provide pathways for research that adequately account for gender roles and relations as a foundation for better research, policy and, ultimately, practice in adoption of climate-resilient strategies. In this part we provide methodological issues underpinning the integration of gender into adoption behaviour. To explore this, the part is divided into three chapters:

Chapter 11 discusses seven steps in which appropriate gender analysis can be conducted for climate-smart intervention.

Chapter 12 summarizes the mediator and moderator variables applicable to gender analysis.

Chapter 13 provides a recap of methodological arguments summarized in previous chapters and brings a practical action research model to frame 'gender mainstreaming' for implementation of climate-resilient agricultural practices.

The book provides relevant research material (including resource guides, tools, websites, etc.) in the Reference Guides and Websites section. The Glossary provides standard definitions of relevant concepts/terminology used in the book.

The book covers the entire process for gender-responsive climate-smart intervention. While many of the resources referenced in the book are relevant across regions, we have used text boxes, case studies, resource lists and relevant literature to highlight examples globally.

1.3.2 Who is this guide for and how should it be used?

The book is planned as a regular book for students (especially upper-level undergraduate and postgraduate), scholars, researchers and practitioners interested in gender and agricultural development. We believe it could be used for supplementary courses for subjects related to gender and development, climate mitigation and environmental studies.

Each section of the book is self-contained. Readers can go directly to those chapters, sections and methods of particular relevance to their work. The reader can use it in development of their research proposal, research design, research article and academic thesis in the following ways:

1. Basic understanding: overview of key gender analysis topics related to climate change and climate change adaptation. This book can be used as an inventory of:
   a. What are the prevailing issues? What projects or policy interventions have happened and have been researched?
   b. What issues are researchable and which ones need to be addressed as early as possible, and how?
   c. What are the existing methodologies and potential methodologies to understand and address these issues?

2. Mainstreaming approach towards gender equality:
   a. Are the research goals disaggregated by gender?
   b. Does the climate change issue affect men and women in different ways?
   c. Are their gender balances in the programmes as well as in the institutions and bodies involved?
   d. The extent to which gender issues are addressed: Are the needs of both men and women being met?
   e. Exploration of social and historical disadvantage faced by men and women in society.

3. Engendering communication: synthesis of gender perspectives in designing and implementing international and national policies. Documentation of climate change adaptation-related gender resources from global review to region-specific examples, and concise case studies.

4. Methods and approaches used for gender inclusion: research approaches, methods and tools used to examine gender inclusion, efficiency and efficacy (coverage, scope, sustainability) of the research.
5. Recommendations and ideas for carrying out the research: identifying gaps, suggesting potential approaches and research issues for entry points and indicators.

The book demonstrates potential solutions for reducing losses due to climate risk, thus the two terms used mostly are 'adoption' and 'adaptation'. Zilberman *et al.* (2012) have discussed the difference between these two terms. According to them:

> Adoption is change in practice or use of technology by economic agents. The adoption literature investigates, among other things, patterns of adoption, the profile of adopters, the timing of adoption, the evolution of the technology once adopted, and impacts on other technologies and prices. Adaptation is a response to a shock (e.g. climate change) and consists of many actions, including adoption.
>
> (ibid.)

The solutions, however, depend on timing, scale and magnitude of adoption, among other features, and can be divided into coping, adaptation and mitigation strategies. These are discussed in Chapter 7 in detail. For climate solutions in general we have used the term 'climate-resilient strategies' (CRS) where the strategies include actions and choices made before and after climate events. Readers are encouraged to remember the definitions.

Notes

[1] Information retrieved from: http://www.fao.org/state-of-food-security-nutrition/en/
[2] Word cloud (or tag clouds), earlier used for a visual representation of words, and now also considered for analysing text data (DePaolo and Wilkinson, 2014).
[3] Information retrieved from climate-related and other emerging risks disclosures, such as assessing financial statement materiality using AASB/IASB Practice Statement 2, p. 6 (accessed 26 May 2020).
[4] 'Socially constructed' means that these are not 'givens' or 'natural' but are constructed or produced by society and as such can be modified and changed (European Institute of Gender Equality, 2013).

References

Adger, W.N., Brown, I. and Surminski, S. (2018) Advances in risk assessment for climate change adaptation policy. *Philosophical Transactions of the Royal Society A* 376, 20180106.
ADPC (2006) *The economic impact of the 26 December 2004 earthquake and Indian Ocean tsunami in Thailand*. Technical paper. Asian Disaster Preparedness Center (ADPC), Disaster Risk Management System Department, Thailand.
Alkire, S., Meinzen-Dick, R., Peterman, A., Quisumbing, A.R., Seymour, G. *et al.* (2013) The women's empowerment in agriculture index. *World Development* 52, 71–91.
Boserup, E. (1970) *Women and Economic Development.* St. Martin's Press, New York.

Brody, A., Demetriades, J. and Esplen, E. (2008) *Gender and Climate Change: Mapping the Linkages: A Scoping Study on Knowledge and Gaps*. BRIDGE, Institute of Development Studies (IDS), UK.

CDP (2015) *The Forgotten 10%: Climate Mitigation in Agricultural Supply Chains*. CDP, New York.

CGIAR (2015) *Definitions of Gender Research for CRP Gender Budgets*. Guidance note. CGIAR Gender and Agriculture Research Network.

CGIAR-IEA (2017) *Evaluation of Gender in CGIAR Research and in the CGIAR Workplace*. Independent Evaluation Arrangement (IEA) of the CGIAR, Rome.

Cheung, W.W., Jones, M.C., Reygondeau, G., Stock, C.A., Lam, V.W. *et al*. (2016) Structural uncertainty in projecting global fisheries catches under climate change. *Ecological Modelling* 325, 57–66.

Dankelman, I. (2002) Climate change: learning from gender analysis and women's experiences of organizing for sustainable development. In: Masika, R. (ed.) *Gender, Development and Climate Change*. Oxfam, London, pp. 21–29.

Delaney, P. and Sharder, E. (2000) *Gender and Post-Disaster Reconstruction: The Case of Hurricane Mitch in Honduras and Nicaragua: Decision Review Draft*. The World Bank: LCSPG/LAC Gender Team.

Demombynes, G. and Kiringai, J. (2011) *The Drought and Food Crisis in the Horn of Africa: Impacts and Proposed Policy Responses for Kenya*. The World Bank. Economic Premises, no. 71.

DePaolo, C.A. and Wilkinson (2014) Get your head into the clouds: using word clouds for analyzing qualitative assessment data. *TechTrends: Linking Research and Practice to Improve Learning* 58(3), 38–44.

Doss, R.C. and Kieran, C. (2014) *Standards for Collecting Sex-Disaggregated Data for Gender Analysis: A Guide for CGIAR Researchers*. CGIAR Gender and Agriculture Research Network. International Food Policy Research Institute, Washington, DC.

ECOSOC (1997) Report of the Economic and Social Council for 1997 (A/52/3, 18 September). Available at: http://www.un.org/womenwatch/daw/csw/GMS.PDF (accessed 23 February 2019).

Esplen, E. (2009) *Gender and Care: Overview Report*. Bridge: Development and Gender, UK.

European Institute of Gender Equality (2013) Review of the Implementation in the EU of Area K of the Beijing Platform for Action: Women and the Environment – Gender Equality and Climate Change. Available at: https://op.europa.eu/en/publication-detail/-/publication/9ef701cd-3c76-48a7-8739-eb1fc126ffa7 (accessed 12 November 2019).

FAO (n.d.). Module 1: Conceptual Framework: Gender Issues and Gender Analysis Approaches. Rome. Available at: http://www.fao.org/climatechange/31734-02e69e384b4ef76bd49477a9650ef4db6.pdf (accessed 3 March 2022).

FAO (2011) The State of Food and Agriculture Women in Agriculture: Closing the Gender Gap for Development 2010–2011. Food and Agriculture Organization of the United Nations, Rome. Available at: http://www.fao.org/publications/sofa/2010-11/en/ (accessed 3 March 2022).

FAO (2018) 2017: The Impact of Disasters and Crises on Agriculture and Food Security. Food and Agriculture Organization of the United Nations (FAO), Rome. Available at http://www.fao.org/3/I8656EN/i8656en.pdf (accessed 10 April 2022).

Hausmann, R., Laura, D.T. and Zahidi, S. (2006) *The Global Gender Gap Report 2006*. The World Economic Forum, Geneva. Available at: http://www.weforum.org/pdf/gendergap/report2006.pdf (accessed 10 April 2022).

IEWM (n.d.) Gender-smart Adaptation to Climate Change. Policy Brief. Institute of Environment and Water Management, Nairobi. Available at: http://iewm.org/documents/cc/Policy%20brief%20%20Gender-smart%20adaptation%20to%20climate%20change.pdf

ILO (1998) ILO/SEAPAT's On-line Gender Learning & Information Module. Unit 1: A Conceptual Framework for Gender Analysis and Planning. www.ilo.org/public/english/region/asro/mdtmanila/training/unit1/groles.htm (accessed 3 March 2022).

IPCC (2007a) *Climate Change 2007: Impacts, Adaptation and Vulnerability* (edited by Parry, M.L., Canziani, O.F., Palutikof, J.P., van der Linden, P.J. and Hanson, C.E.). Contribution of Working Group II to the Fourth Assessment Report of the Intergovernmental Panel on Climate Change. Cambridge University Press, Cambridge, UK.

IPCC (2007b) Pachauri, R.K. and Reisinger, A. (eds) *Climate Change 2007: Synthesis Report*. Contribution of Working Groups I, II and III to the Fourth Assessment Report of the Intergovernmental Panel on Climate Change. IPCC, Geneva, Switzerland.

IPCC (2012) Glossary of terms. In: Field, C.B., Barros, V., Stocker, T.F., Qin, D., Dokken, D.J. *et al*. (eds) *Managing the Risks of Extreme Events and Disasters to Advance Climate Change Adaptation*. A Special Report of Working Groups I and II of the Intergovernmental Panel on Climate Change (IPCC). Cambridge University Press, Cambridge, UK, and New York, pp. 555–564.

IPCC (2019) Annex 1: Glossary. In: *IPCC Special Report on Climate Change, Desertification, Land Degradation, Sustainable Land Management, Food Security, and Greenhouse Gas Fluxes in Terrestrial Ecosystems*. IPCC, Geneva, Switzerland, p. 28.

Jutting, J.P., Morrisson, C., Dayton-Johnson, J. and Drechsler, D. (2008) Measuring gender (in) equality: the OECD Gender, Institutions and Development Data Base. *Journal of Human Development* 9(1), 65–86.

Kabeer, N. (1999) Resources, agency, achievements: reflections on the measurement of women's empowerment. *Development and Change* 30(3), 435–464.

Knox, J., Hess, T., Deccache, A. and Wheeler, T. (2012) Climate change impacts on crop productivity in Africa and South Asia. *Environmental Research Letters* 7, 034032.

Lambrou, Y. and Nelson, S. (2010) *Farmers in a Changing Climate: Does Gender Matter?* Food and Agricultural Organization of the United Nations (FAO), Rome.

Lopez-Claros, A. and Zahidi, S. (2005) *Women's Empowerment: Measuring the Global Gender Gap*. World Economic Forum, Geneva, Switzerland.

Mehar, M. and McDougall, C. (2018) *Methods and Tools for Gender Analysis in FISH: A Preliminary Consolidation and Reference Guide*. CGIAR Research Program on Fish Agri-Food Systems, Penang, Malaysia.

Mehar, M., Mittal, S. and Prasad, N. (2016) Farmers' coping strategies for climate shock: Is it differentiated by gender? *Journal of Rural Studies* 44, 123–131.

Okin, S.M. (1989) *Gender, Justice and the Family*. Basic Books, New York.

Pauw, K., Thurlow, J., Bachu, M. and van Seventer, D.E. (2012) The economic costs of extreme events: a hydrometeorological CGE analysis for Malawi. *Environmental and Development Economics* 16, 177–198.

Reeves, H. and Baden, S. (2000) *Gender and Development: Concepts and Definitions*. BRIDGE Report 55. Institute of Development Studies, Brighton, UK.

Ribot, J.C., Magalhes, A.R. and Panagides (1996) *Climate Variability, Climate Change, and Social Vulnerability in the Semi-Arid Tropics*. Cambridge University Press, UK.

Rojas-Downing, M.M., Nejadhashemi, A.P., Harrigan, T. and Woznicki, S.A. (2017) Climate change and livestock: impacts, adaptation, and mitigation. *Climate Risk Management* 16, 145–163.

Sibhatu, K.T. and Qaim, M. (2017) Rural food security, subsistence agriculture, and seasonality. *PLOS ONE* 12(10), e0186406.

Stevanović, M., Popp, A., Lotze-Campe, H., Dietrich, J.-P., Muller, C. *et al*. (2016) The impact of high-end climate change on agricultural welfare. *Science Advances* 2(8), e1501452.

UNDP (1995) *Human Development Report*. Oxford University Press, New York.

United Nations (1992) *United Nations Framework Convention on Climate Change*. FCCC/INFORMAL/84, United Nations, New York.

Wunder, S., Noack, F. and Angelsen, A. (2018) Climate, crops, and forests: a pan-tropical analysis of household income generation. *Environmental and Development Economics* 23, 279–297.

Yohannes, H. (2016) A review on relationship between climate change and agriculture. *Journal of Earth Science and Climate Change* 7(2), 335.

Zilberman, D., Zhao, J. and Heiman, A. (2012) Adoption versus adaptation, with emphasis on climate change. *The Annual Review of Resource Economics* 4, 27–53.

2 Gender Matters in Agriculture and Climate Change

Abstract

The nature of women's agricultural work is, to a greater extent than for men, casual and often under-represented. Climate change hazards increase gender inequalities. Often women are viewed as victims of climate change, though they are also contributors to it. Men and women, within their capacities, also perform different activities to reduce the effects of climate risks. This chapter presents the theoretical underpinnings surrounding the discourse of gender and climate change using interpretative paradigms. The literature and the theoretical background provide useful conceptual links, which are used in constructing the methodologies of the next chapters of the book.

2.1 Introduction

Gender is one of the important dimensions to understand the impact of climate change at both the micro and macro level. Culture and social roots mean women and men live in a society at a point in time, though there are differences in the resources and opportunities provided for them. The traditional power structures are often obscured in favour of one gender. The obscured power creates power imbalances between women and men leading to inequitable access and control of resources and opportunities, including the exclusion of one gender from the decision-/policy-making process or leadership.[1] Such gender differences exacerbate vulnerability to climate risks. The differentiation in the roles of women and men gives rise to different vulnerabilities and affects the ability of the different genders to respond to climate risks and solutions. However, though there is no common view between men and women about the impact of climate change, many social realities exist due to varying human experience, including people's knowledge, views, interpretations and experiences.

This chapter attempts to construct a simple framework using an interpretative paradigm to analyse what and how gender roles and relations change in the

DOI: 10.1079/9781789249910.0002

context of coping strategies and in the context of climate events in agriculture. The framework considers three main concepts: (i) climate variation; (ii) gender and power dynamics; and (iii) agriculture. The climate variation, as defined above, considers four main parameters: temperature, precipitation, extreme events and seasonal pattern. Climate variation, as defined earlier, considers changes in historical climate patterns arising from natural and/or man-made causes. More specifically, climate variation can be gauged from the parameters of variation in temperature, precipitation, seasonal pattern and occurrence of extreme events. The concept of 'climate variation' is relevant to understand to what extent people are concerned about changes in weather and their impact on agriculture and livelihoods. Broadly, the gender and power dynamics are shaped from cultural constructions which contribute to changes in perception, upbringing and thinking. The parameters defining gender-differentiating power dynamics are: (i) biological differences; (ii) roles and responsibilities; (iii) capacity and capability; (iv) human responses; and (v) institutional and policy dimension.

2.2 Interpretative Paradigm

The link between gender and climate variation in agriculture is more influenced by the subjective experiences of individuals or society. The perception of humans of climate variation cannot be explained entirely by any scientific paradigm where hypotheses are pre-structured and, to some extent, results are known. In such contexts there are no 'correct' or 'incorrect' theories and no 'correct' knowledge, so it can be better explained with an interpretative paradigm[2] (Walsham, 1995). The interpretative paradigm is concerned with understanding the context and gives the research a greater scope to address issues of influence and impact. In this chapter, an interpretative paradigm is applied to enhance the understanding of 'why' and 'how' farmers' gender identity, roles and relations trajectories to climate change are created. 'The movement of understanding is constantly from the whole to the part and back to the whole' (Gadamer, 1976, p. 117), so it is a circular relationship. Hence, these trajectories are developed through meaning and understanding of the related concepts developed socially and experientially from the literature review, the author's judgement and following principles of interpretative paradigm (Box 2.1).

2.3 Conceptual Framework

Though there are pieces of evidence suggesting the differential impact of climate variation in agriculture on male and female farmers, the direct link between climate variation and gender in the context of agriculture is not clear. This weak linkage is represented by a broken line that connects gender and climate variation in Fig. 2.1. The associated path connecting the link is agriculture. Total productivity and sustainability of agriculture are influenced by farmers' access to technology, information and their practices and beliefs, for optimal utilization of their limited means. The differential access to resources, information and priorities leads to gender-differentiated impacts.

Box 2.1. Summary of principles for interpretative paradigm. (From Klein and Myers, 1999, p. 72)

Principles	Descriptions
The fundamental principle of the Hermeneutic Circle	This principle of human understanding is fundamental to all the other principles of interpretative paradigm. This principle suggests that all human understanding is achieved by iterating between considering the interdependent meaning of parts and the whole that they form.
The principle of contextualization	This principle requires clear reflections of the social and historical background of the research setting to ensure the intended researcher is able to see how the current situation under investigation emerged.
The principle of interaction between the researchers and the subjects	This principle requires clear reflections on how the data or research materials were constructed through the interaction between researchers and respondents to the study.
The principle of abstraction and generalization	This principle requires relating the idiographic details revealed by the data interpretation through the application of principles one and two to theoretical, general concepts that describe the nature of human understanding and social action.
The principle of dialogical reasoning	This principle requires understanding of potential contradistinctions between the theoretical preconceptions guiding the research design and actual findings with subsequent cycles of revision.
The principle of multiple interpretations	This principle requires understanding of potential differences in interpretation among the respondents as are typically expressed in multiple narratives or stories of the same sequence of events under study.

The links between gender and climate variation in the context of agriculture are bi-directional. Changes in climate events affect agriculture through a reduction in yield, increase in the occurrence of pests or crop disease and changes in time of activities; for example, changes in wintertime could result in a change in harvesting or sowing days of crops. All this affects farmers' livelihood and work activity in different ways. For example, women and girls often have to travel long distances after a drought to collect water. Shreds of evidence suggest a decrease in yield and thereby total availability of food consumption in a household basket usually has been complemented by a reduction in meals for a female. This exposes them to health issues. Male farmers are usually seen to migrate in search of non-farm opportunities outside their village domain, implicitly putting a burden of responsibility on women. Moreover, mostly in rural areas, women are less likely to get an education as compared to men; they are less mobile and do more household-related activities, and they have less control over the family assets.

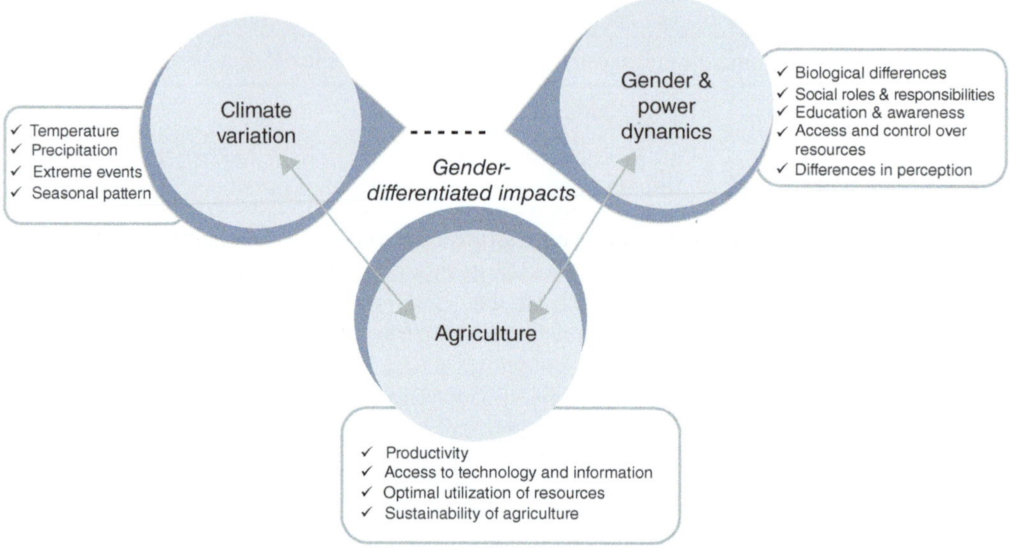

Fig. 2.1. Conceptual framework to define the linkage between gender and climate change. (From Mehar, 2014)

The other direction of gender-differentiated impacts in the context of climate events in agriculture can be viewed through understanding male and female coping strategies. The adoption of any strategy is often influenced by the priorities and options available for women and men for coping with these strategies. Constraints on women's mobility and inequitable access to assets and resources hinder their ability to decide on various matters and thus they have few choices of coping strategies available. The assets and resources may include access to land, knowledge, technology, power in decision making, education, healthcare and food, etc. This inequality leads to gender-differentiated impacts in adverse climate situations.

Since this chapter focuses on answering questions of 'why' and 'how' men's and women's perceptions and actions are linked in climate variation, the association route between agriculture and climate change, agriculture and gender, and gender and climate change is important to identify. From here onwards we discuss these paths in detail.

2.3.1 Agriculture and climate change

Climate change and agriculture are interrelated. Several studies (Sinha and Swaminathan, 1991; Kumar and Parikh, 2001; Lambrou and Piana, 2006; Cline, 2007) have recognized that the agricultural sector is one of the major contributors to climate change as well as being adversely affected by it. Between 2006 and 2016, 26% of the total damage and losses caused by climate-related disasters in 53 developing countries were absorbed by the agricultural sector (FAO, 2018). Of these, 83% of all damage and loss was caused by drought (ibid., p. 16). By percentage share, damage and loss were highest for crops (49%), followed by livestock (36%), fisheries (3%), forestry (4%) and unspecified (8%) (ibid., p. 17). The report stated:

One of the most direct ways in which natural disasters affect the sector is through reduced production. This results in direct economic loss to farmers, which can cascade along the entire value chain, affecting agricultural growth and rural livelihoods.

The trajectories of the interrelationship between climate change and agriculture are depicted in Fig. 2.2.

Agriculture affects climate change. On one side, sustainable agricultural practices and natural carbon sequestration in soils act as an effective carbon sink. On the other, in the past few decades, increasing agricultural area through deforestation has increased greenhouse gas (GHG) emissions due to excessive use of fertilizers and pesticides; ruminant digestions (cattle, sheep, goats, etc.); rice cultivation by the rice-transplanting method; fuel use; burning of crop residue after harvesting; and industrialization of agriculture, and they have acted as major contributors to carbon footprint and climate change. The majority of the world's greenhouse gas emissions come from land use (30%), with the largest share from crop production (64%) (De Pinto, 2018). In another study (IPCC, 2007), direct GHG emissions from cultivation and livestock are estimated to account for about 10–12% of total GHG emissions.

Climate change affects agriculture. Climate and extreme climate events affect agriculture in diverse ways and on varying scales. Cline (2007) has assessed the impact of climate change on agriculture in 116 countries. He found a global reduction in agricultural productivity (output per hectare) of 16% without carbon fertilization, and a reduction of 3% with carbon fertilization effects. He further mentioned that damage will generally be greater in countries located closer to the equator, such as India. However, the effect of climate change on agriculture was both positive and negative depending on crop type, regional or geographical location and status of socio-economic development. Climate change affects agricultural production in two ways: (i) direct effects on plants from changes in temperature, precipitation, etc.; and (ii) indirect effects, which can further be divided into two parts:

- through the loss of natural resources – increases in weed, insects, pests and disease; changes in soil moisture; impact on biodiversity with loss of genetic diversity;
- through affecting the quantity and quality of food and thus affecting the health, food security and livelihood of farmers.

Direct effect. Changes in climatic factors (e.g. temperature, rainfall) affect agricultural productivity through physiological changes in crops (Chakraborty *et al.*, 2000). After a disaster, the water that is needed for agriculture is no longer available. Also, quite often the sources available are polluted and cannot be used directly for daily needs (drinking, washing, cooking) or for agricultural activities.

Indirect effect. One of the most dramatic effects of climate change is the depletion of natural resources, such as water availability. For example, farmers in Bangladesh reported a marked decrease in depth of the main canals and an increase in elevation of arable land due to sedimentation (Kabir *et al.*, 2017).

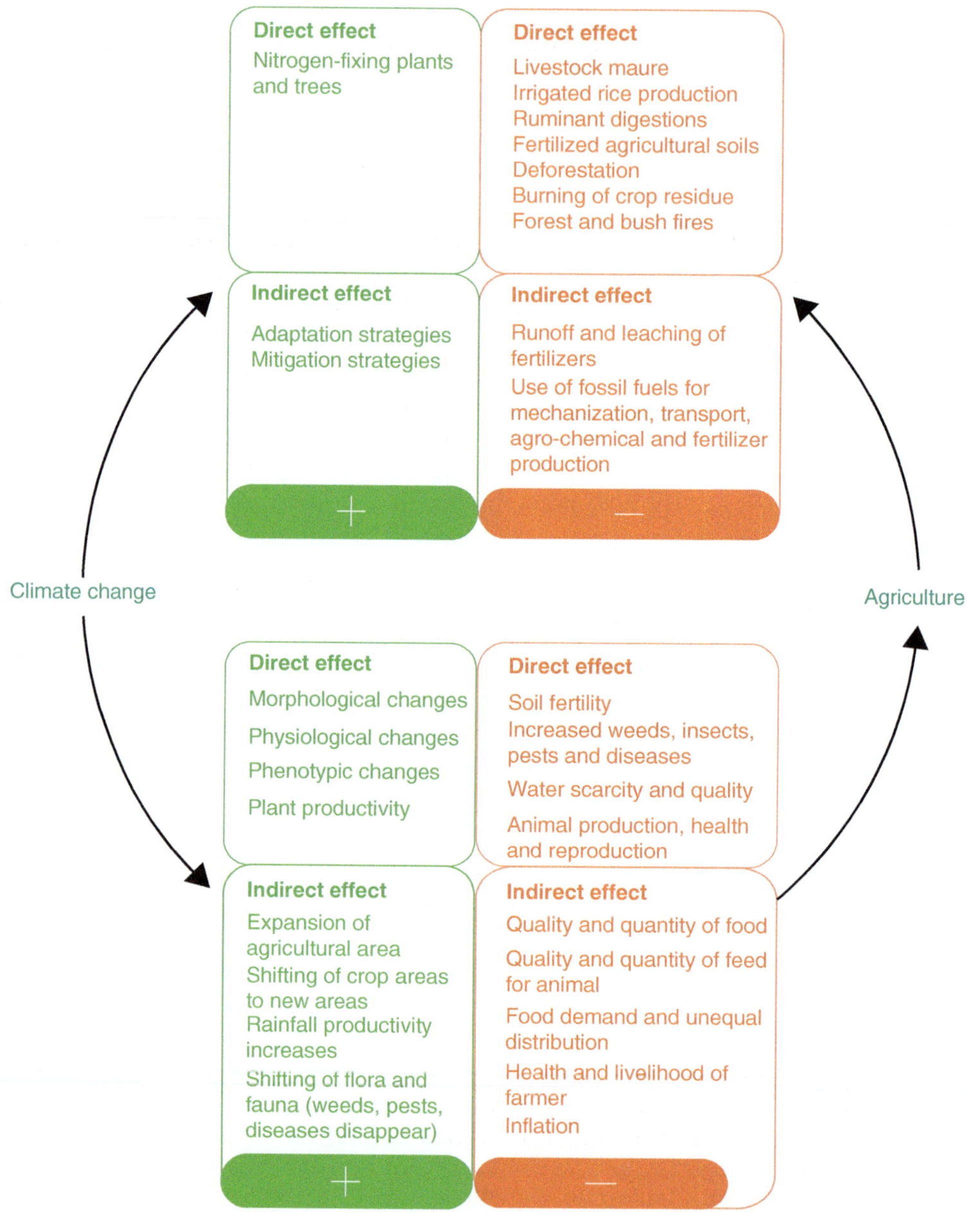

Fig. 2.2. Trajectories of climate change and agriculture. (From: the author)

Climate extremes have also increased the species and intensity of weeds, insects and pest outbreaks. In a few crops, climate change has also caused plant disease. The decline in soil fertility is observed by both naturally induced (especially in coastal areas where increased flooding raises the level of soil salinity) and human-induced climate effects (e.g. by applying excessive chemicals in agriculture). This will cause the plants to suffer from nutrient deficiency. To mitigate climate risk, several agricultural practices and cropping patterns have been adopted. Due to low rainfall, methods such as deep cultivation and drip irrigation or rainwater harvesting are promoted.

A reduction in the quantity and quality of agricultural production as a result of climate change further threatens the world's and countries' food and nutrition security programmes. For farmers, this threat is an additional layer of existing constraints which include land-tenure insecurity, natural resource degradation, financial deficiencies, inadequate services and infrastructure provision, and weak institutional arrangements. Such a shock will push households further into poverty; their livelihoods are affected due to farmers moving to off-farm work opportunities and a reduction in expenditure on food, health and education.

2.3.2 Agriculture and gender

> 'If we invest in women, they can feed the world' (Bill and Melinda Gates Foundation[3]).

Gender goes beyond biological differences; it refers to the social attributes and opportunities that are associated with being a male or a female, relationships between women and men and girls and boys, as well as the relations between women and between men (UN, 2001). These attributes, opportunities and relationships are socially constructed or produced by society, and as such can be modified or changed (ibid.).

Global evidence shows women work as much as, or more than, men in the agricultural sector. Globally, about 43% of workers engaged in agricultural activities are women (Akter *et al.*, 2017), and across Asian and African countries about half of all agricultural workers are women (Agarwal, 2015). However, their contribution is often overlooked and described as 'invisible farmers'. Kassie *et al.* (2015) argue that increasing women's empowerment by extending their ability to make decisions and take advantage of opportunities is essential for broad-based agricultural development in low-income countries.

The gender wheel (Fig. 2.3) was first used by Parker *et al.* (1995) as a tool for training in gender awareness. It portrays the dynamics of gender roles and relations at the household and individual level. The first two dimensions of the wheel are *gender roles* and *division of labour* and are often constructed by culture and social ideologies. According to UNESCO, gender *division of labour* is 'the result of how each society divides work among men and women according to what is considered suitable or appropriate to each gender' (UNESCO, 2003). 'In most societies, there are clear patterns of "women's work" and "men's work," both in the household and in the wider community – and cultural explanations of why this should be so' (Koskinen, 2008, p. 166). Women usually have key responsibilities for household activities and productive agricultural tasks (e.g. weeding, transplanting, water and fodder collection, processing and drying fish, livestock

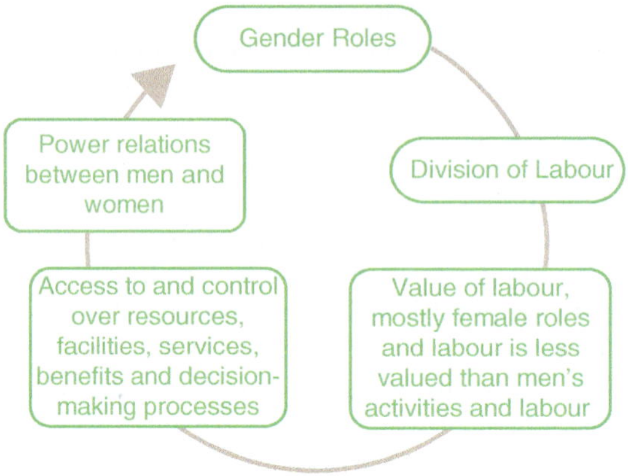

Fig. 2.3. Gender wheel. (From: the author)

rearing) while men are largely involved in land preparation and marketing. In forestry, men dominate in harvesting timber and hunting while women dominate wild plant collection and wood fuel collection (Sunderland *et al.*, 2014). Kiptot and Franzel's (2012) review of agroforestry studies in Africa from a gender perspective document several promising approaches to improving women's benefits from agroforestry, although the authors have also acknowledged ritual and cultural prohibitions on women's engagement in agroforestry (ibid.). Similar to agriculture and forestry, fisheries and livestock are gendered spaces. In fisheries, women are involved in the processing and selling of marine fish and take care of ponds for aquaculture. Women and girls in developing countries are often the primary collectors, users and managers of wood for fuel and water for domestic consumption and spend a disproportionate amount of time on these tasks, thereby experiencing 'time poverty'. Low rainfall and drought have a secondary effect, like girls dropping out of school (Kanda *et al.*, 2017).

Boserup (1970) argued that, due to physical strength, men, on average, have an advantage in farming compared with women, resulting in a specialization of production along gender lines in societies that traditionally practised plough cultivation. A review of gender division in agricultural production showed that the jobs done by women are time-consuming. However, more economic value (and therefore income) is put on men's jobs, which are viewed as more labour-intensive (Trung, 2013). Binswanger-Mkhize and D'Souza (2012) also mention that, relative to men, women continue to have lower real wage rates and lower average real earnings in both agriculture and non-agriculture. Other strong values and assumptions include women's mobility.

Limited mobility is also a constraint for female farmers' *access to and control over resources*. Constraints on mobility, whether cultural or because of time burdens, may hinder women from leaving the household to join community-based groups, interacting with extension agents, purchasing inputs from dealers, or seeking information and services outside of the village (Fletschner and Mesbah, 2011). The 'perception bias' towards men may be particularly strong in South

Asia where women in male-headed households do not typically manage their plots as they do in most parts of Africa (Peterman *et al.*, 2010; Quisumbing and Pandolfelli, 2010); but it is also important to acknowledge that while the specific nature of gender relations varies among societies (as societies are not homogeneous), the general pattern is that women have less personal autonomy, fewer resources at their disposal and limited influence over the decision-making processes that shape their societies and their lives (OECD, 2012, p. 86). A growing body of literature has documented gender bias in the use of agricultural inputs (Udry, 1996; Chen *et al.*, 2011). Worldwide, female farmers have less access to land (World Bank, 2011) and studies on ownership and control of land have found gender inequalities in Asia (Estudillo *et al.*, 2001; Agarwal, 2003). Agarwal (2003) emphasizes the social perceptions in society as the reason why men are the legitimate claimants of land. Women farmers often have greater difficulty in obtaining fertilizers and water, particularly in African and Asian countries (Kinkingninhoun-Mêdagbé *et al.*, 2008; Agarwal, 2015; Kassie *et al.*, 2015).

Control over resources affects the *decision-making process and power dimension*. The decision making around which crop, which technology, which seed and which coping strategies to adopt is strongly influenced by the four types of power (Fig. 2.4). Migrated men are found to remotely control farm decisions. Studies have shown mixed findings regarding women's power to make farming decisions in a household where men are migrated (Hadi, 2001; Sinha *et al.*, 2012; Choithani, 2019). Women are often restricted from extracting information from extension agents and other farmers. They are also restricted from engaging in the agricultural community due to mobility constraints.

The above description of the gender wheel is presented through the lens of cultural and social norms, though the cycle is not static, as over time and space

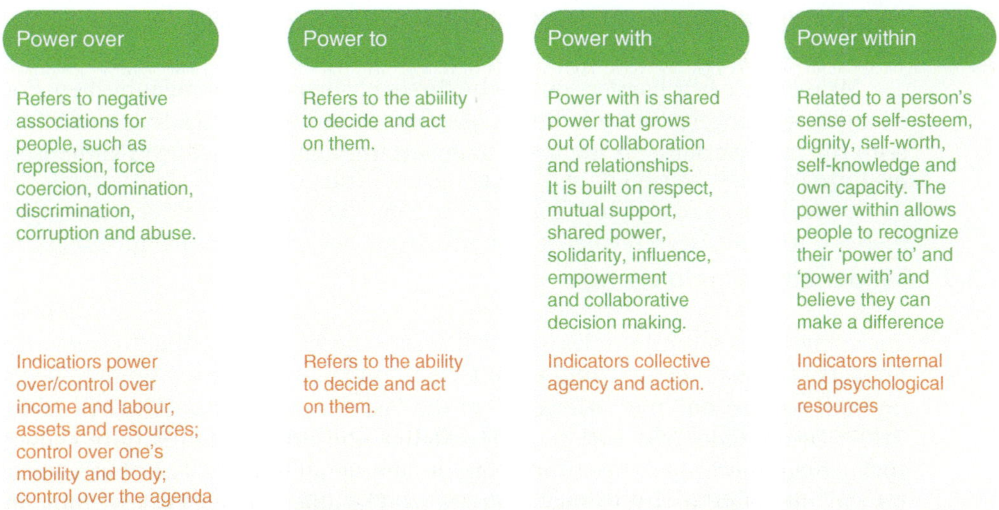

Fig. 2.4. Power types. (From: the author)

the attributes, patterns and explanations embedded in the gender wheel may differ. The other two dimensions by which understanding of the gender wheel can be grasped are 'capital' and 'institutional barriers'. They are also influenced by socio-cultural norms. Gendered roles, gendered actions and effects of gender roles and actions are determined and discriminated in the access to resources.

Given that institutions and new technologies are influenced by local context, the ways in which men and women participate in them, adopt them and are influenced by them vary. For example, the new technology is often designed targeting men, for example the tractor. The introduction of the tractor endorsed the existing labour division of men and women in land preparation. The technology, though, results in increased leisure time for men but does not affect pre-existing or predefined activities of women such as weeding and transplanting (in the case of rice). Another example of technologies designed in favour of men as well as climate change is zero tillage, though the latter results in increased weeding, so increasing the burden on women.

The institutional barriers are prevalent in the entire value chain cycle. For example, women are viewed as livestock-raisers and growers of vegetables in the home garden, and sell them locally. However, with commercialization, the milk and vegetables sold to retail markets, and the income generated, are controlled by the husband. Another example is maize crop pounding, which is done by women, but men take and sell them at the market, and the mills are also often operated by men.

Another example of the institutional barrier is bias in targeted information provided to men and women. In 1999, the Bangladesh cyclone warning 'was transmitted by men to men in public spaces, but rarely communicated to the rest of the family, and as many women are not allowed to leave the house without a male relative, they perished waiting for their relatives to return home and take them to a safe place' (Wong, 2012, p. 15). Similarly, in Peru, early warning messages about the arrival of El Niño (which is an oceanographic phenomenon associated with warmer-than-normal water temperature (Carlowicz and Schollaert, 2017) were only transmitted to the fishermen (UNISDR-UNDP-IUCN, 2009).

Women also face barriers due to the threat of sexual harassment by men. For example, a cultural practice in Kenya of *jaboya* means women trading sex for fish with the men working on the boats to ensure they get the best fish (Nathenson *et al.*, 2017).

2.3.3 Gender and climate change

For many years, several studies have considered the assumption that the negative impacts of climate change and the efforts to mitigate them have similar effects on both women and men. However, in the last decade, studies (Mitchell *et al.*, 2007; Demetriades and Esplen, 2010; Mehar and Prasad, 2015) have shown that women and men experience climate change differently and that gender inequalities often worsen women's situations. The differences vary depending on their farming type, household economic status (e.g. ownership of agricultural and non-agricultural assets), social status, etc. These differences lead to different reasons and choices to address climate risk. Fig. 2.5 summarizes the gender and climate change interconnections.

Mortality. The immediate and most apparent evidence is the difference in numbers of casualties between men and women after a natural hazard. A report by UNDP (2010), from a sample of 141 countries, from 1981 to 2002, found that natural disasters (and their subsequent impacts) kill more women than men, on average, or kill women at an earlier age than men. More than 70% of people who died in the 2004 Asian tsunami were women. Similarly, hurricane Katrina, which hit New Orleans in 2005, predominantly affected poor African-Americans, especially women. Another study by CARE in Myanmar reported 61% of females (out of 130,000 people) were dead or missing after a cyclone. Similarly, Pradhan *et al.* (2007) documented in their study that out of 40,000 households that experienced floods in 1993 in Nepal, about 80% (out of which 60% were women and girls) suffered flood-related fatalities. This may be attributed to their inability to participate in social and educational events, which leads to them being disadvantaged in skills like swimming (MacDonald, 2005). Drought intensifies the vulnerability and marginalization of women owing to loss of livelihood (Singh *et al.*, 2013). In this sense, women and girls are more likely to die from natural disasters (Williams, 2015). Table 2.1 summarizes the female mortality evidence for major climate-related disasters.

Effect on health. Rising temperatures may increase risks, especially among subgroups of pregnant women or women with multiple pregnancies in low and middle income countries (LMIC) as these subgroups have reduced physiological ability to respond to heat (Basu *et al.*, 2017; Chersich *et al.*, 2020). Most gender-differentiated health issues are due to gender-discriminatory norms, practices and biases that determine division of labour. For example, women have the roles of cooking, collecting drinking water (in some places collecting water for agricultural purposes), collecting firewood for fuel, etc. Water collection tasks, due to walking long distances (especially after droughts) and carrying heavy containers on the head or the hip, cause damage to the spine and the lower back. Over 4 million

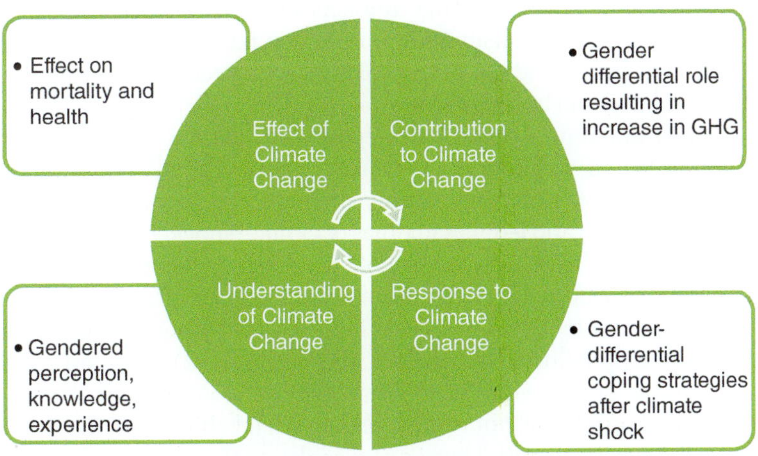

Fig. 2.5. Gender and climate change connections. (From: the author)

Table 2.1. Female mortality evident in major climate-related disasters. (From Pradhan *et al.*, 2007; UNDP, 2010; Glemarec *et al.*, 2016)

Year	Disaster	Country/region	Female mortality reported (%)
1991	Cyclone OB2	Bangladesh	90
1993	Flood	Nepal	60*
2004	Tsunami	Indonesia	77
2004	Tsunami	India (Tamil Nadu)	73
2008	Cyclone Nargis	Myanmar	61
2009	Tsunami	Tonga and Samoa	70
2014	Floods	Solomon Island	96*
2015	Earthquake	Nepal	55

*Reported for women and children

people each year die prematurely due to illnesses caused by indoor air pollution, primarily from smoke produced while cooking with solid fuel (UNDP, 2010).

Contribution to GHG. Agriculture is a major contributor to global warming, and the contributing factors can be divided into men- and women-led activities. Men perform activities such as residue burning, high-tech use of agricultural technologies, excessive use of chemicals (pesticides, insecticide, weedicide, etc.), timber harvesting and excessive fishery capture. Women are viewed as natural custodians of livestock and food security (including cooking). Emission of methane from livestock as well as from cooking stoves is a major contributor to climate change.

Gendered perception, knowledge and experience. Both men and women perceive climate as changing and becoming dangerous, though there are differences in their perception bounded by their knowledge and experience. The overall majority of studies found people's experience, knowledge and perception reveal that weather (average temperature, frequent flood and drought, late rainfall) is changing fast, especially during recent decades, though there are variations among these parameters of context and scale. It is also evident that women and men experience climate change impacts differently due to their socially constructed roles and responsibilities, on the one hand, and their adaptive capacity and cost, self-belief, and the risk of taking on new technologies on the other. This is discussed in detail in Chapter 4.

Gender-differential coping strategies after climate shock. The gender roles, division of labour, differences in resource rights, and differences in priorities and preferences, all of which are highly dynamic, lead to differences in capability approach of males and females in times of climate shock. Entire families often renegotiate their roles to achieve household food security after climate change occurs. Families often shift to eat cheap grains while women often report eating less. For example, women in Nepal reported shifting from eating rice to eating millet and chino after crop failures (Gentle and Maraseni, 2012). Men, on the other hand, often found alternative off-farm jobs or migrated to find work. That work is often insecure and unreliable, but women are often left behind with the responsibility for crops, livestock and family. There is also evidence of men controlling crop decisions from a distance and sending remittances. This is discussed in detail in Chapter 5.

Many scholars and practitioners have reviewed trends on the gender dimension and documented that numerous guidelines and climate-smart practice documents have already been produced (Bernier *et al.*, 2015; FAO, 2017; Kristjanson *et al.*, 2017). This will be discussed more in Chapters 3 and 5. (Refer also to the Reference Guides and Websites section.)

In summary, gender differences manifest themselves in the effects (in terms of mortality and health issues) of climate change, in vulnerability and capacity to adapt to this change. Without a proper understanding of these dimensions and the complexity of issues involved in them, design and implementation of effective and feasible gender-sensitive policies and support mechanisms are not possible.

2.4 Summary

The conceptual framework in this chapter discusses routes of climate and gender connection in agriculture. Climatic events adversely affect the total yield of agriculture and thereby reduce the food basket and consumption of farm households, especially women. Also, there are gender differences in adapting to different coping strategies in such situations. Male farmers opt for strategies like out-migration in search of additional employment, selling off land, etc., while females are bound by their capacity as well as capability and do not have such options at their disposal; thus they are forced to choose options which adversely affect them, such as reducing food intake.

The discussion above and the conceptual framework did not explicitly discuss the feminist and masculinity discourse of climate change. The literature on these is limited.

Notes

[1] The interpretative paradigm points out that positivism, in its first attempt to model the social after the natural sciences, fails to see that, unlike nature, social reality exists only insofar as lay members create that reality in meaningful interaction (Fuchs, 1992, p. 198). But the point of interpretative paradigm is that we must first understand social worlds from within before we develop scientific models and explanations (ibid., p. 205).

[2] From two flagship reports which focused on gender and agriculture (Ley, 2012).

References

Agarwal, B. (2003) Gender and land rights revisited: exploring new prospects via the state, family and market. *Journal of Agrarian Change* 3(1-2), 184–224.

Agarwal, B. (2015) Food security, productivity, and gender inequality. In: Herring, R.J. (ed.) *The Oxford Handbook of Food, Politics and Society* (ch. 11). Oxford University Press, Oxford, UK.

Akter, S., Rutsaert, P., Luis, J., Htwe, N.M., San, S.S. *et al.* (2017) Women's empowerment and gender equity in agriculture: a different perspective from Southeast Asia. *Food Policy* 69, 270–279.

Basu, R., Chen, H., Li, D.K. and Avalos, L.A. (2017) The impact of maternal factors on the association between temperature and preterm delivery. *Environmental Research* 154, 109–14.

Bernier, Q., Meinzen-Dick, R.S., Kristjanson, P., Haglund, E., Kovarik, C., Bryan, E. and Ringler, C. (2015) *Gender and Institutional Aspects of Climate Smart Agricultural Practices: Evidence from Kenya*. CCAFS Working Paper No. 79, Copenhagen. CGIAR Research Program on Climate Change, Agriculture and Food Security (CCAFS).

Binswanger-Mkhize, H.P. and D'Souza, A. (2012) Structural transformation and agricultural productivity in India. In: Fugile, K., Wang, S.L. and Bal, E. (eds) *Productivity Growth in Agriculture: An International Perspective*. USDA/CAB International, Wallingford, UK.

Boserup, E. (1970) *Women and Economic Development*. St. Martins Press, New York.

Carlowicz, M. and Schollaert, S. (2017) El Niño: Pacific wind and current changes bring warm, wild weather. Available at: https://earthobservatory.nasa.gov/features/ElNino (accessed July 2019).

Chakraborty, S., Tiedemann, A.V. and Teng, P.S. (2000) Climate change: potential impact on plant diseases. *Environmental Pollution* 108(3), 317–326.

Chen, S.E., Bhagowalia, P. and Shively, G. (2011) Input choices in agriculture: Is there a gender bias? *World Development* 39(4), 561–568.

Chersich, M.F., Pham, M.D., Areal, A., Haghighi, M.M., Manyuchi, A. *et al.* (2020) Associations between high temperatures in pregnancy and risk of preterm birth, low birth weight, and stillbirths: systematic review and meta-analysis. *BMJ* 371, m3811.

Choithani, C. (2019) Gendered livelihoods: migrating men, left-behind women and household food security in India. *Gender, Place & Culture* 1–22.

Cline, W.R. (2007) *Global Warming and Agriculture: Impact Estimates by Country*. Center for Global Development and Peterson Institute for International Economics, Washington, DC.

De Pinto, A. (2018) The global effects of widespread adoption of climate smart agriculture. In: 30th International Conference of Agricultural Economists, 28 July–2 August, Vancouver, BC. Available at: http://ageconsearch.umn.edu/record/277524 (accessed 5 March 2022).

Demetriades, J. and Esplen, E. (2010) The gender dimensions of poverty and climate change adaptation. In: Mearns, R. and Norton, A. (eds) *Social Dimensions of Climate Change: Equity and Vulnerability in a Warming World*. The International Bank for Reconstruction and Development/The World Bank, Washington, DC.

Estudillo, J.P., Quisumbing, A.R. and Otsuka, K. (2001) Gender differences in land inheritance and schooling investments in the rural Philippines. *Land Economics* 77(1), 130–143.

FAO (2017) *The Role of Gender in Climate-Smart Agriculture. Module C-6 in Climate-Smart Agriculture Sourcebook*. Food and Agriculture Organization of the United Nations, Rome. Available at: http://www.fao.org/climate-smart-agriculture-sourcebook/enabling-frameworks/module-c6-gender/c6-overview/en/ (accessed 5 March 2022).

FAO (2018) *2017: The Impact of Disasters and Crises on Agriculture and Food Security*. Food and Agriculture Organization of the United Nations, Rome. Available at: http://www.fao.org/3/I8656EN/i8656en.pdf (accessed 5 March 2022).

Fletschner, D. and Mesbah, D. (2011) Gender disparity in access to information: Do spouses share what they know? *World Development* 39(8), 1422–1433.

Fuchs, S. (1992) *The Professional Quest for Truth: A Social Theory of Science and Knowledge*. SUNY Press, Albany, USA.

Gadamer, H.G. (1976) The historicity of understanding. In: Connerton, P. (ed.) *Critical Sociology, Selected Readings*. Penguin Books, Harmondsworth, UK, pp. 117–133.

Gentle, P. and Maraseni, T.N. (2012) Climate change, poverty and livelihoods: adaptation practices by rural mountain communities in Nepal. *Environmental Science & Policy* 21, 24–34.

Glemarec, Y., Qayum, S. and Olshanskaya, M. (2016) *Leveraging Co-benefits between Gender Equality and Climate Action for Sustainable Development: Mainstreaming Gender Considerations in Climate Change Projects*. UN Women, New York.

Hadi, A. (2001) International migration and the change of women's position among the left-behind in rural Bangladesh. *International Journal of Population Geography* 7(1), 53–61.

IPCC (2007) *Climate Change 2007: The Physical Science Basis. Contribution of Working Group I to the Fourth Assessment Report of the Intergovernmental Panel on Climate Change* (edited by Solomon, S., Qin, D., Manning, M., Chen, Z., Marquis, M., Averyt, K., Tignor, M.M.B., and Miller, H.L.). Cambridge University Press, Cambridge, UK and New York, NY, USA.

Kabir, J.M., Cramb, R., Alauddin, M., Roth, C. and Crimp, S. (2017) *Farmers' Perception of and Responses to Environmental Changes in Southwest Coastal Bangladesh. Asia Pacific Viewpoint* 58(3).

Kanda, A., Murongazvombo, M. and Ncube, F. (2017) Adapting household water use in rural Zimbabwe. *International Journal of Environmental Studies* 74(3), 471–485.

Kassie, M., Stage, J., Teklewold, H. and Erenstein, O. (2015) Gendered food security in rural Malawi: Why is women's food security status lower? *Food Security* 7(6), 1299–1320.

Kinkingninhoun-Mêdagbé, F.M., Diagne, A., Simtowe, F., Agboh-Noameshie, A.R. and Adégbola, P.Y. (2008) Gender discrimination and its impact on income, productivity, and technical efficiency: evidence from Benin. *Agriculture and Human Values* 27(1), 57–69.

Kiptot, E. and Franzel, S. (2012) Gender and agroforestry in Africa: a review of women's participation. *Agroforestry Systems* 84(1), 35–58.

Klein, H.K. and Myers, M.D. (1999) *A set of principles for conducting and evaluating interpretive field studies in information* systems. *MIS Quarterly* 23(1), 67–94.

Koskinen, P. (2008) 'Asian values', gender, and culture-specific development. In: Kingsbury, D. and Avonius, L. (eds) *Human Rights in Asia: A Reassessment of the Asian Value Debates*. Palgrave Macmillian, New York, pp. 159–181.

Kristjanson, P., Bryan, E., Bernier, Q., Twyman, J., Meinzen-Dick, R. *et al.* (2017) Addressing gender in agricultural research for development in the face of a changing climate: Where are we and where should we be going? *International Journal of Agricultural Sustainability* 15(5), 482–500.

Kumar, K.S.K. and Parikh, J. (2001) Indian agriculture and climate sensitivity. *Global Environmental Change* 11, 147–154.

Lambrou, Y. and Piana, G. (2006) *Gender: The Missing Component of the Response to Climate Change*. Food and Agriculture Organization of the United Nations, Rome.

Ley, H. (2012) If we invest in women, they can feed the world. Available at: http://www.impatientoptimists.org/Posts/2012/03/Sharing-a-Global-Agenda-for-Womens-Empowerment-in-Agriculture (accessed 14 August 2014).

MacDonald, R. (2005) How women were affected by the tsunami: a perspective from Oxfam. *PLOS Medicine* 2(6), e178.

Mehar, M. (2014) Gender specific impacts of climate variation in agriculture. Unpublished MPhil thesis, Department of Economics, School of Social Science, Indira Gandhi National Open University, India.

Mehar, M. and Prasad, N. (2015) Gender participation in rural farm household decision making: a case of Vaishali district, Bihar. *Indian Journal of Economics and Development* 11(3).

Mitchell, T., Tanner, T. and Lussier, K. (2007) *We Know What We Need: South Asian Women Speak Out on Climate Change Adaptation*. Action Aid and Institute of Development Studies (IDS) at the University of Sussex, Brighton, UK.

Nathenson, P., Slater, S., Higdon, P., Aldinger, C. and Ostheimer, E. (2017) No sex for fish: empowering women to promote health and economic opportunity in a localized place in Kenya. *Health Promotion International* 32(5), 800–807.

OECD (2012) *Poverty Reduction and Pro-Poor Growth: The Role of Empowerment*. OECD Publishing.

Parker, R., Lonzano, I.A. and Messner, L.A. (1995) *Gender Relations Analysis: A Guide for Trainers*. Save the Children, New York.

Peterman, A., Behrman, J. and Quisumbing, A. (2010) *A Review of Empirical Evidence on Gender Differences in Non-Land Agricultural Inputs, Technology, and Services in Developing Countries*. International Food Policy Research Institute.

Pradhan, E.K., West, K.P., Katz, J., LeClerq, S.C., Khatry, S.K. *et al.* (2007) Risk of flood-related mortality in Nepal. *Disasters* 31(1), 57–70.

Quisumbing, A.R. and Pandolfelli, L. (2010) Promising approaches to address the needs of poor female farmers: resources, constraints, and interventions. *World Development* 38(4), 581–592.

Singh, R., Feroze, S.M. and Ray, L.I.P. (2013) Effects of drought on livelihoods and gender roles: a case study of Meghalaya. Research note. *Indian Journal of Gender Studies* 20(3), 453–467.

Sinha, B., Jha, S. and Negi, N.S. (2012) Migration and empowerment: the experience of women in households in India where migration of a husband has occurred. *Journal of Gender Studies* 21(1), 61–76.

Sinha, S.K. and Swaminathan, M.S. (1991) Deforestation, climate change and sustainable nutrition security: a case study of India. *Climate Change* 19, 201–209.

Sunderland, T., Achdiawan, R., Angelsen, A., Babigumira, R., Ickowitz, A. *et al.* (2014) Challenging perceptions about men, women, and forest product use: a global comparative study. *World Development* 64, S56–S66.

Trung, P.T. (2013) Climate change and its gendered impacts on agriculture in Vietnam. *International Journal of Development and Sustainability* 2(1), 52–62.

Udry, C. (1996) Gender, agricultural production and the theory of the household. *Journal of Political Economy* 104(5), 1010–1046.

UN (2001) *Gender Mainstreaming: Strategy for Promoting Gender Equality*. Office of the Special Adviser on Gender Issues and Advancement of Women. Available at: http://www.un.org/womenwatch/osagi/pdf/factsheet1.pdf (accessed 10 April 2022).

UNDP (2010) *Gender, Climate Change and Community-Based Adaptation*. UNDP, New York.

UNESCO (2003) UNESCO's gender mainstreaming implementation framework: baseline definitions of key concepts and terms. Available at: http://www.unesco.org/new/fileadmin/MULTIMEDIA/HQ/BSP/GENDER/PDF/1.%20Baseline%20Definitions%20of%20key%20gender-related%20concepts.pdf (accessed 5 March 2022).

UNISDR-UNDP-IUCN (2009) Making disaster risk reduction gender-sensitive. UNISDR, UNDP and IUCN, Geneva, Switzerland. Available at: http://www.preventionweb.net/files/9922_MakingDisasterRiskReductionGenderSe.pdf (accessed 5 March 2022).

Walsham, G. (1995) Interpretive case studies in IS research: nature and method. *European Journal of Information Systems* 4, 74–81.

Williams, M. (2015) *Gender and Climate Change Financing: Coming Out of the Margin*. Routledge, London.

Wong, S. (2012) Gender relations, migration and urban social capital in Hong Kong. In: Streich, G.W. and Lewandowski, J.D. (eds) *Urban Social Capital: Civil Society and City Life*. Routledge, London, pp. 265–276.

World Bank (2011) *World Development Report 2012: Gender, Equality and Development*. The World Bank, Washington, DC.

3 Gender-inclusive Climate Policies

Abstract

Gender considerations have been given attention in climate change policy regarding agriculture in the last two decades. This paper outlines key past policies and sets out an argument for gender mainstreaming of climate policy. It also lays the institutional mechanism and structural barriers in policy implications partly reflected by inappropriate framing and understanding of gender by policymakers and implementers.

3.1 Introduction

> We are aware that gender roles in climate change adaptation are fluid, negotiated, and even contested; what matters most for gender justice under climate change is how women and men can equally access and control decision-making processes and challenge discriminatory institutional practices.
>
> Bee *et al.* (2012, p. 104)

In the realm of the gender wheel components, quite significant work on gender and climate change has been carried out and has highlighted the need for gender-inclusive policies with special attention to women. The integration of gender into climate change strategies is important to:

- ensure gender equality and human rights, particularly women's rights and climate justice;
- ensure consumption smoothing after climate shocks, as women are considered the natural custodians of food security of the household and men as the prime breadwinners; and
- adapt effective climate-resilient strategies (CRS) and low-carbon development efforts.

DOI: 10.1079/9781789249910.0003

The need for considering gender as an important category in environmental policy and practice was seeded with the clarion call of 'No climate justice without gender justice' at the Bali COP conference in 2007. Yet while some progress has been made towards climate justice (see Reference Guides and Websites), until recently, very few CRS and low-carbon developments had considered a comprehensive gender-responsive approach.

This chapter seeks to investigate the provision of gender in international climate change policies over the years. The degree of gender integration in policies is discussed along a continuum from 'gender exploitative' to 'gender transformative' adaptation plans. Lastly, the chapter shares the common misconceptions, misinterpretations and misunderstandings of 'gender' concepts in climate policies.

3.2 Evolution of Gender Inclusiveness in Climate Policies in Major International Agreements

Numerous international environmental agreements and climate commitments which include gender concerns in climate policymaking have been enacted at transnational levels to address climate change, as listed in Table 3.1. The view to include gender concerns in climate policy started in the 1970s. The instance that gained attention was the Chipko movement in the Garhwal division of Uttar Pradesh state (India) when women hugged trees to prevent logging; though there are mixed views: some represent it as women's concern for conservation, thus a good example of their environmental commitment (Leach, 2007); others consider it not as evidence of women's closeness to nature but as a struggle to save the natural resources they use in daily life (i.e. wood for fuel, trees for food or medicines) (Jackson, 1993). In 1972, a seminar – Women and the Environment Crisis – was organized by the Environmental Liaison Centre International (ELCI). In 1985, at the UN Third World Congress on Women in Nairobi, attention to women's environmental issues was brought to an official symposium and for the first time at a global level. In 1985, 1500 women from 87 countries met to present women's voices about the deterioration of the environment worldwide. This effort resulted in:

- the development of the Women's Environment & Development Organization (WEDO) in 1990; and
- the development of Agenda 21, which was the first milestone for integration of women and gender issues into the sustainable development agenda.

Agenda 21 was the outcome of the UN Conference on Environment and Development (UNCED), also known as the *'Earth Summit'*, and was held in Rio de Janeiro. According to the Rio Declaration, 'countries should develop a gender-sensitive database for the analysis of the structural linkages between gender relations, environment and development' (UN, 1992, p. 273). After this, most of the following UN Conferences of Parties (COP) included some discussion on gender, the word 'women' appearing for the first time, officially, at COP7 (in Marrakesh in 2001). Gender equality references were integrated at COP16 (in Cancun in 2010);

Table 3.1. Key gender-specific climate policies and commitments.

Year	Agreement	Descriptions and strategic action points
1990	Women's Environment & Development Organization (WEDO)	With the aim to bring women's voices to the environmental agenda, WEDO was established by former US Congresswoman Bella Abzug and feminist activist Mim Kelber. In 2016, WEDO developed a Gender Climate Tracker App, which aims to highlight key gender entry points to climate policies.
1992	United Nations Conference on Environment & Development, Rio	Global action for women towards sustainable and equitable development (Agenda 21).
1995	Beijing Declaration and Platform for Action Fourth World Conference on Women	The declaration called for the active involvement of women in environmental decision making at all levels: – involve women actively in environmental decision making at all levels; – integrate gender concerns and perspectives in policies and programmes for sustainable development; and – strengthen or establish mechanisms at the national, regional and international level to assess the impact of development and environmental policies on women.
2001	UNFCCC (COP7), Marrakesh	One of the first documents to include a gender perspective was improving the participation of women in the representation of parties.
2002	World Summit on Sustainable Development (WSSD), Johannesburg	The summit adopted a set of policy recommendations that recognized women's role in disaster reduction, response and recovery.
2005	Hyogo Framework for Action (HFA) World Conference on Disaster risk Reduction, Kobe	Called for efforts towards building the resilience of nations and communities to disasters 2005–15 (HFA) and Commission on the Status of Women resolutions (in the 49th session) on gender equality and the empowerment of women in natural disasters.
2007	Stockholm Forum of Disaster Reduction and Recovery, Stockholm	The Forum recognized the need for long-term, gender-inclusive engagement to address underlying vulnerability and poverty in order to sustain progress towards the MDGs (Millennium Development Goals).
2007	UNFCCC (COP 13), Bali	Establishment of Women's Action Group (Gender CC).
2009	UNFCCC (COP 16), Mexico	Establishment of the Troika+ of Women Leaders on Women and Climate Change. The Troika+ played a crucial role to the Doha agreement (2012) which resulted in promoting gender balance and improving the participation of women in UNFCCC negotiations and in the representation of Parties in bodies established pursuant to the Convention or the Kyoto Protocol. In same year, Women and Gender Constituency (WGC) was formed as one of the nine stakeholder groups of the UNFCCC, made up of 15 women's and environmental civil society organizations and networks.

Continued

Table 3.1. Continued.

Year	Agreement	Descriptions and strategic action points
2010	UNFCCC (COP16), Cancun	References to gender were included as an important part of climate action.
2011	UNFCCC (COP17), Durban	UNFCC adopted a decision that Guidelines for National Adaptation Plans (NAPs) should be country-driven, participatory, with a gender-sensitive approach.
2012	UNFCCC (COP18), Doha	The decision on 'Promoting gender balance and improving the participation of women in UNFCCC negotiations and in the representation of Parties in bodies established pursuant to the Convention or the Kyoto Protocol'.
2012	UNFCCC (COP18), Doha	Reiterated and strengthened the above decision at COP7.
2013	UNFCCC (COP19), Warsaw	Emphasis on actions on collection and management of relevant data including gender-disaggregated data.
2014	UNFCCC (COP20), Lima	The Parties agreed on a Lima Work Programme on Gender to advance gender balance and promote gender sensitivity in developing and implementing climate policy. This includes the appointment of a senior gender experts to the UNFCCC and gender trainings for delegates. National Adaptation Plan (NAP) global network was established to support gender-supportive approaches.
2015	Sendai Framework for Disaster Risk Reduction (DRR)	Adopted at the World Conference on DRR as a successor instrument to the Hyogo Framework for Action and calls for stronger gender, age and disability perspectives in policies and practices; promotion of women's leadership; women's participation in DRR; resourcing and implementing gender-sensitive DRR; and adequate capacity-building measures for women's empowerment and livelihoods.
2015	UNFCCC (COP21), Paris	Gender equality, empowerment of women and intergenerational equity.
2015	Sustainable Development Goals (SDGs) (The 2030 Agenda)	There is a strong gender focus in SDG 5 (Achieve gender equality and empower all women and girls) – with nine specific targets and gender-sensitive targets for most other SDGs. Also, SDG 10 (Reducing inequality) is relevant in this context. All SDGs, including SDG 13 (Climate action), have important gender-equality implications and require a gender-sensitive approach to meet the 2030 targets.
2016	UNFCCC (COP22), Morocco	Invites parties to appoint and provide support for a national gender focal point for climate negotiations, implementation and monitoring.

Continued

Table 3.1. Continued.

Year	Agreement	Descriptions and strategic action points
2017	UNFCCC (COP23), Bonn	UNFCCC Parties adopted a new roadmap to incorporate gender equality and women's empowerment in climate change discourse and actions. The creation of a Gender Action Plan was agreed upon by the Parties to bolster the role of women in climate action.
2018	UNFCCC (COP24), Katowice	Parties adopted the Paris Implementation Guidelines with explicit reference to the importance of gender responsiveness and gender balance, taking into account social aspects and including traditional/indigenous/local knowledge in several areas: the Nationally Determined Contributions (NDCs), adaptation communication, finance, technology framework, transparency framework, and compliance.
2019	UNFCCC (COP25), Chile	Parties agreed on a **five-year enhanced Lima Work Programme on Gender and its Gender Action Plan (GAP)**.
2019	INDC (Intended Nationally Determined Contributions) Partnership Gender Strategy	In accordance with the SDG 2030 agenda and Paris Agreement, nations signal their contribution through NDC to cut their greenhouse gases. NDC gender strategy emphasized gender equality and women's empowerment as a high priority for the partnership.
2021	UNFCCC (COP26), Glasgow	A dedicated day tagged 'Gender Day' decided to recognize and celebrate gender equality.

the decision to promote gender balance in UNFCCC and its delegations at COP18 (in Doha in 2012); the decision to establish a two-year work programme for promoting gender balance at COP20 (in Lima in 2014) (Box 3.1); the extended Lima Work Programme on Gender for three years at COP22 (in Marrakesh, Morocco, in 2016); and the five-year enhanced Lima Work Programme at COP26 (in Chile in 2019). There is no denying that 'The United Nations is formally committed to gender mainstreaming in all policies and programmes, and that should include policymaking processes relating to climate change' (Hemmati and Röhr, 2009). In addition, UN sustainable development goals (2015) emphasize adopting gender-sensitive approaches in all SDGs including SDG-13 (climate action) to meet the 2030 targets. Dazé and Dekens (2017) present a working framework for gender-responsive National Adaptation Plans (NAPs) and the NDC partnership gender strategy aims at facilitating the gender-mainstreaming process for the development and implementation of nations' commitments to cutting their own greenhouse gases (NDC, 2019).

These commitments and efforts remind us that climate change, initially, was viewed as a technical and scientific problem and not people-orientated. This view makes it difficult to include gender and social inclusion in climate

Box 3.1. The LIMA Work Programme on Gender.

In COP20 (Decision 18/CP.22), UNFCCC launched the LIMA Work Programme on Gender to enhance previous UNFCCC efforts on gender. The key actions to emerge from this were (UNFCCC, 2015):

1. efforts to build a common understanding in terms of the concepts related to gender mainstreaming in the context of gender-responsive climate policies; and
2. the decision to strengthen the existing work on gender balance in the thematic priority area and encourage parties to support:
   - training and awareness-raising for female and male delegates on issues related to gender balance and climate change;
   - building the skills and capacity of their female delegates to effectively participate in UNFCCC meetings via training on, *inter alia*, negotiation skills, drafting of legal language and strategic communication;
   - deciding to establish a two-year work programme for promoting gender balance and achieving gender-responsive climate policy;
   - deciding to clarify the meaning of the term 'gender-responsive climate policy' from an implementation perspective. For this, parties called for two workshops with a focus on:
     - mitigation action and technology development and transfer (June 2015); and
     - adaptation, capacity building and training delegates on gender issues (May 2016).

The Paris Agreement (2015) especially calls for all parties to prepare and communicate INDCs aimed at monitoring emissions and mitigation efforts of nations. An analysis of the 162 submitted INDCs shows that only 40% (65) have explicitly made 'gender' or 'women' references in their national plans. Out of these 65, 13 mention the words exclusively in their adaptation sections and three mention them exclusively in their mitigation sections (IUCN and USAID, 2016). Only half of these 65 INDCs identify gender as cross-cutting policy priority to integrating gender into all climate change actions, strategies and policies. Twelve countries specifically point out the special vulnerability of women to climate change. Seven countries mention gender concerns in relation to agriculture, four intend to strengthen the resilience of women to disasters and only three explicitly mention the engagement of women in the planning process.

At COP22 (UNFCCC, 2016), parties decided to continue and enhance the Lima Work Programme on Gender for three years and invite parties to continue to assist:

- training and awareness-raising for female and male delegates on issues related to gender balance and climate change; and
- building the skills and capacity of their female delegates to participate effectively in UNFCCC meetings through training on, *inter alia*, negotiation skills, the drafting of legal documents and strategic communication.

Even in the most recent COP26 (in Chile in 2019), parties agreed on a five-year enhanced Lima Work Programme (also known as the Paris Agreement) on gender and gender action.

Indeed, the Paris Agreement, to date, is considered to be the most important milestone in gender and climate change policy.

change and emission-reduction policies. This view changed with the acceptance of the climate justice concept invented in the 1980s, though it was only accepted seriously after the UN climate change conference (COP) in Copenhagen in 2017.[1]

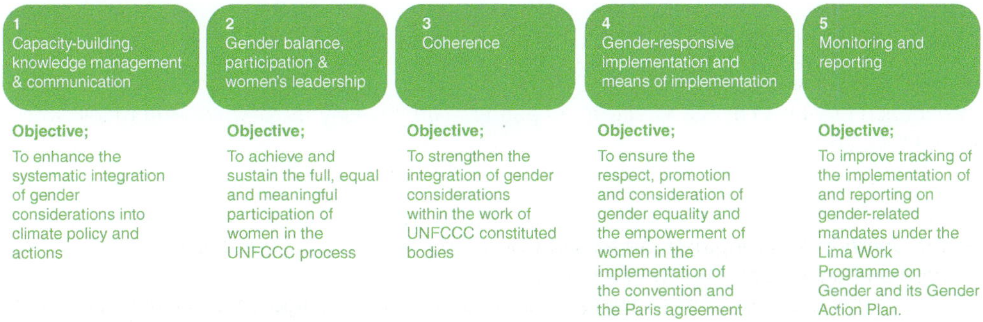

Fig. 3.1. Priority areas of the Gender Action Plan (GAP). (https://unfccc.int/topics/gender/workstreams/the-gender-action-plan [accessed 1 March 2020])

The most recent programme is a Gender Action Plan for five years (2018–2023), focusing on adaptation, mitigation and the means of implementation (Fig. 3.1). Aguilar *et al.* (2015) have summarized the action plans of 16 national governments who adopted ccGAP (climate change GAP). The 2013 Environment Gender Index showed that India, Pakistan and Bangladesh had the highest performance as compared to developed countries (IUCN, 2013).

In compliance with the above international frameworks, national governments draft their climate change policies and laws. It is also important to note that international donors have a key role in adopting a gender and climate change agenda; though, as stated by Otzelberger (2011), 'most of them [climate related programme donors] lack effective strategies for systematic integration of gender in their adaptation and mitigation work'. Other stakeholders (including civil society, the private sector, etc.) have recently been emerging to take up low-carbon initiatives with a social and gender dimension. Other programmes also follow a 'do no harm' approach which means 'that new practices or changes in farming systems for climate change adaptation, mitigation, or climate-smart agriculture (CSA) will not compromise food security and create greater social inequality' (FAO and World Bank, 2017, p. 9).

Only a few institutions and organizations working with environment and energy programmes have initiated comprehensive gender-mainstreaming policies, while others have failed to do so or have resulted in the creation of standalone women's divisions or programmes within institutions (UNDP, 2007; SERVIR-Mekong, 2015).

3.2.1 Gender equality continuum tool

There are different approaches for including gender in the formulation and implementation of climate policies. At the broadest level, these can be explained using the gender equality continuum model developed by IGWG and USAID (IGWG and USAID, 2017). The model describes programmes and policies on a spectrum from gender-blind to gender-aware, with the latter including gender-accommodative and gender-transformative (Fig. 3.2). We have provided relevant examples for gender approaches on the spectrum which depict how the aspects of gender and climate are recognized.

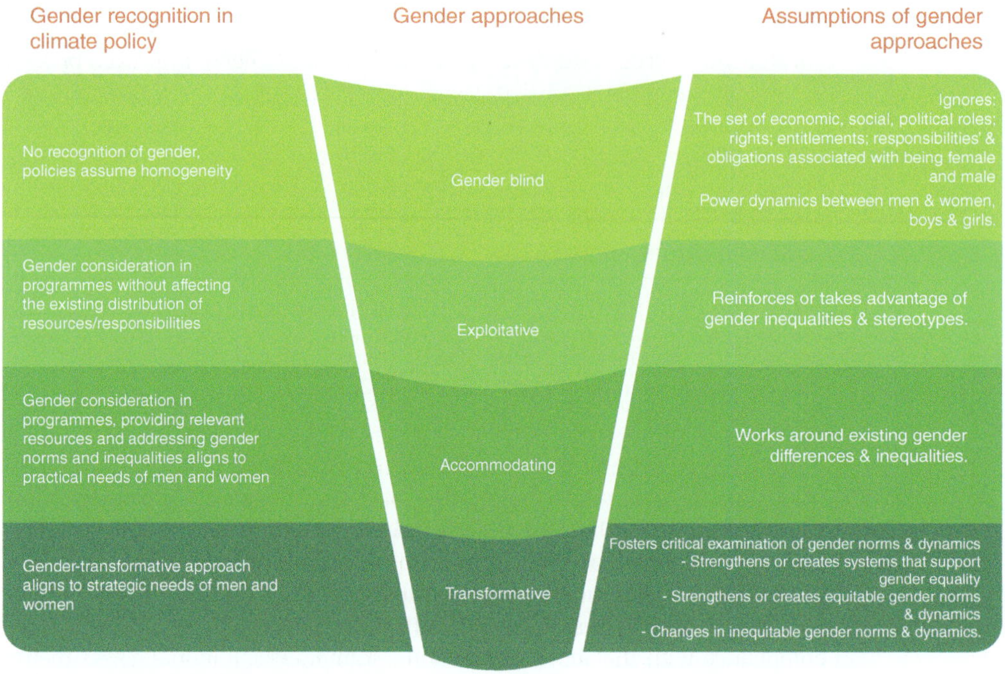

Fig. 3.2. Gender inclusion in climate change policy. (From: the author)

Gender blindness. Ignoring the different roles, responsibilities and capabilities of men, women, boys and girls, and the social processes that determine these. Gender-blind policies and programmes are based on male-centric experiences as the 'norm' and on the assumption that everyone affected by them has the same needs and preferences (Kabeer 2003, p. 243).

> EXAMPLE
> Several climate-responsive technologies (such as zero-tillage) are built in keeping with men's needs (based on their role in agriculture) and capability (physical strength) to use them.

Gender-aware. Includes policies and programmes with the gender dimension. It can be further divided into three categories on the basis of the degree of gender integration. Gender-aware policies, research or interventions consider gender and depend on to what intensity they are included in programmes in which women were viewed as vulnerable, virtuous or active agents to adapting climate solutions. The intensity can be understood from the three sub-categories under gender-aware as follows:

- **Gender-exploitative** programmes explicitly acknowledge the disadvantage faced by women and men and provide them with resources based on their-different experiences and needs. Gender inequalities are not examined or addressed here. The policy, research or interventions follow the 'do no harm' approach and target practical needs, but within the existing distribution of resources and responsibilities.

EXAMPLE

The introduction of women-friendly biogas cookstoves for domestic energy, tools for maize crop-pounding (octagonal maize sheller[2]), improved practices for home/kitchen/garden, to improve time management and reduce the possibility of poverty faced by women. Such activities also have other benefits, e.g. maize crop equipment save women's hands from drying and getting small cuts. Similarly, the use of biogas stoves reduces women's exposure to smoke-related health hazards in addition to reducing drudgery in collecting wood for fuel.

- **Gender-accommodating** or gender-sensitive programmes recognize the need for and provide males and females with resources to address the existing gender norms and inequities. These resources include information, training, equipment and finance relevant to implementation of climate-adaptive strategies. However, these programmes do not challenge the gender roles and norms. The policy, research or interventions consider women as 'virtuous'.

EXAMPLE

Development and provision of women-friendly equipment such as Conoweeder for reducing drudgery in weeding activity. With varying objectives, several initiatives have created women-friendly groups where they directly or indirectly receive support for agricultural activities. For example, microfinance saving groups are quite popular in Bangladesh, especially for fisherwomen in coastal regions. Roof-rainwater harvesting or other small harvesting structures for kitchen gardens or small farms are another example.

- **Gender-transformative** programmes address structural barriers to gender equality involving rethinking roles and opportunities for women and men beyond the social and cultural beliefs. Also known as gender-redistributive, this approach considers women as active agents for adopting climate solutions. Hence, policy, research and intervention focus on building women's capacities and access to resources or addressing strategic needs of men and women.

EXAMPLE

In a flood-prone area, government and other initiatives have tried to provide swimming training to women. In most of these places, women are not allowed to move outside the home, to learn such skills and interact with another adult male. So the initiatives are required to consider practices and programmes that can reduce these barriers. FAO's initiative of gender-sensitive damage assessment and rehabilitation planning process following the 2005 Pakistan earthquake overcame conservative traditions in the area and resulted not only in increased women's membership in watershed management committees but also increased income, confidence and decision-making power (FAO, 2013). In another example of farmer-led and gender-responsive technological innovation, women in Honduras were found to redesign eco-stoves and develop improved agroforestry management systems (Waters-Bayer *et al.*, 2015). The authors stated that 'these women innovators were also found to become more outspoken and active in the community beyond farming activities'.

The embedded thought in the above discussion is that gender-transformation presence assures the existence of gender accommodation and gender exploitation. A function of all these three gender-aware categories can be considered as gender-mainstreaming. Gender-mainstreaming refers to the promotion of gender equality by ensuring that the different needs, experiences and capabilities of all genders are considered when conducting research projects and in making policy recommendations. Gender can be integrated into various programmes and policies at varying degrees. The way and intensity of their inclusion, addressing social norms, make a difference to which category of gender-aware they belong to. For example, the cooking stove, as discussed above, can be gender-exploitative (considering women are the prime cooks), gender accommodating (with provision of reasonable stove price) and gender-transformative (if designed with a participatory approach, or convincing the society (both men and women) to use it). However, the important point to consider is that these technologies must be designed considering the needs and preferences of men and women.

3.2.2 Institutional mechanisms and structural barriers in policy implementation

Despite the growing evidence of international and national creation and implementation of gender inclusion in climate policies, scholars have claimed that gender concerns and voices continue to be the missing piece of the response to climate change (Nelson *et al.*, 2002; Lambrou and Piana, 2006; Hemmati and Röhr, 2009; Alston, 2010; Williams, 2016). For instance, Alston (2010) mentioned 'policies and practices designed to address and shape mitigation and adaptation strategies have failed to incorporate gender mainstreaming'. Hemmati and Röhr (2009) argued the reasons for failure are 'gaps in gender-sensitive data and knowledge about the links between gender justice and climate change; and the lack of participation of women and gender experts in climate-related negotiations'. A related concern is the gender gaps in implementation. Typically, planning is made at the macro level, i.e. international, national, regional/provincial/state; however, their effectiveness depends on whether they reach the individual directly as well as to what extent they have adjusted the planning to local needs and preferences. The issues at meso and micro level are missed or ignored during the implementation process (Fig. 3.3).

For effective policies, collective efforts in governance, decision making and participation are essential. They demand the inclusion of a gender lens throughout the iterative cycle of planning, implementation, monitoring and evaluation (Dazé and Dekens, 2017). This process also includes recognition of gender differences in adaptation needs, opportunities and capacity, and institutional arrangements for equitable access to resources and opportunities such as financial resources. 'As a way of contesting women's marginalisation from environmental politics, feminist scholars have urged not just governments, but NGOs and other institutions' (Ruiz and Vallejo, 2019), including local authorities to facilitate gender inclusion in more sustainable livelihood activities, practices and strategies to address environmental degradation.

O'Neill *et al.* (2010) examined the nature of participants of the UN conference (COP15) in Copenhagen (COP in 2009) on the grounds of disciplines, geography and gender. The article documented that 'an epistemological hierarchy exists in

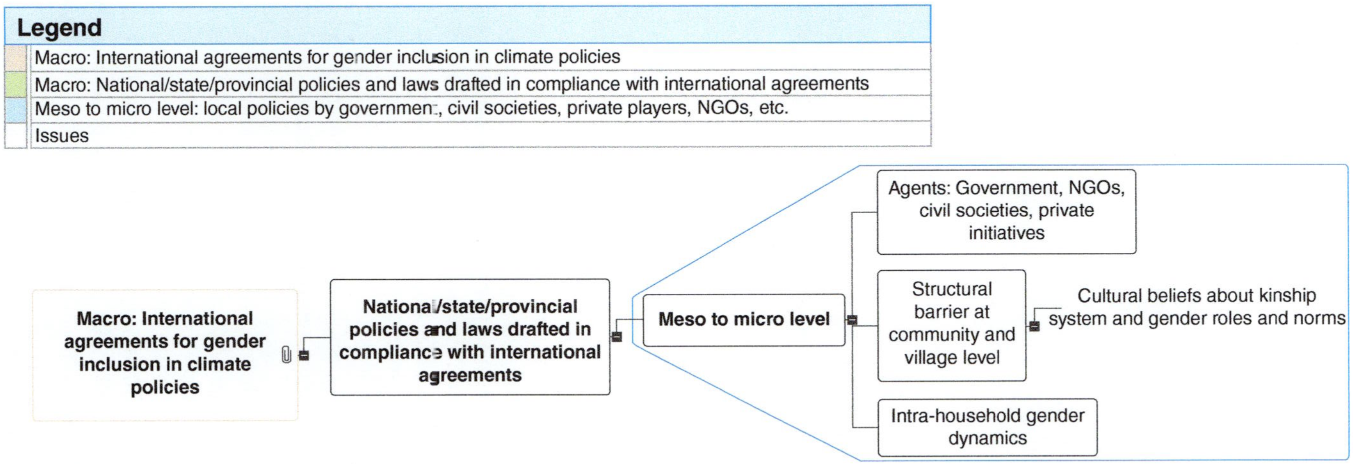

Fig. 3.3. Planning paradigm. (From: the author, using Mindview software)

the framing of climate change [to] disproportionately influence the representation of climate change as primarily an environmental issue'. The author further mentioned the influence is high by developed country participants and male participants. Less than a third of the first authors of presentations were female in the conference. Further, comparing disciplines with a gender lens, less than 20% of presentations of geosciences were from females compared to 40% within social sciences.

Internationally, women are represented in the planning of policies after the decision of participation of women in governmental delegations at UNFCCC COPs in 2001. Thereafter, women representation has improved in science, but we are still far from gender parity (Box 3.2). Different nations have also made mandatory provision of women staff. However, at meso to micro level, women participations are not that essential. This is another reason for loopholes in planning. The next section will discuss in detail the common misconceptions and misinterpretation which leads to gender-biased inclusion and outcomes.

Gay-Antaki (2020) has discussed how 'the meanings of gender changes between negotiating spaces, across moments, and through different bodies that are, in turn, supported or opposed by institutional structures of the COP'. The author found a lack of an intersectional lens in the gender and climate change debate. The author also discussed domination of white, western bodies in the discussions.

Box 3.2. Women in science.

One aspect of feminist science involves ensuring representation for women and gender minorities within science. According to UNESCO, approximately 30% of the world's researchers are women.[3] In 2001, UNFCCC parties agreed on the first text of gender equality and the promotion of women's participation in decision making. In 2012, UNFCCC passed another decision to promote gender balance and women's participation. Since then, UNFCCC has tracked its gender balance. As of 2021, 33% of the members of most bodies under the Convention were women, with the representation varying from 10% to 63% across constituted bodies.[4] According to the Intergovernmental Panel on Climate Change (IPCC), the representation of women scientists went from 5% in 1990 to 20% in 2018 (Gay-Antaki and Liverman, 2018). Other statistics provided by Prebble *et al.* (2015) are that women accounted for 12% of leading environmental-sector ministries (including environment, water, agriculture, forestry, etc.) in an analysis of 193 countries. The author also found that 35% of countries analysed did not have a single female environmental minister. Mavisakalyan and Tarverdi (2019) showed that female representation in parliament results in lower carbon dioxide emissions.

The other aspect is that women's representation is important, but that does not mean gender equality in the field; climate policy processes should go beyond the numerical representation of women to create active mechanisms to express opinions, take initiatives and influence decisions (Huyer *et al.*, 2015). In her analysis of progress in gender equality at the Paris Agreement, Huyer (2016) mentioned that women's representation in UNFCCC bodies is still far from gender parity. Pohlhaus (2017, p. 18) argued that the reason gender has been overlooked in climate research and policy for so long is because of a lack of women scientists, negotiators and members of the UNFCC Secretariat.

Men, equally, need to encourage and actively promote women's voices. One such opportunity is provided by the International Gender Champions Network Platform,[5] which brings men and women together to break down gender barriers and make gender equality a working reality in their spheres of influence.

3.2.3 Misconceptions and concerns for gender inclusiveness

There are many common misconceptions and concerns about gender and gender equality in climate policies at global and local level. These misconceptions often affect the actions of targeted populations and the way programmes are developed and implemented. If these misconceptions are not recognized or dealt with at the start of climate change initiatives, they can have a limiting effect on the project, affecting the ability of men and women to contribute their skills and expertise to strengthen the resilience of all members of society.

1. **Climate change is a scientific matter, so it has nothing to do with gender issues.** Climate change-related policies originally viewed the subject as technically- and scientifically-orientated with no influence on people's lives.
2. **Planning paradigm.** One of the biggest challenges is the transparency and development of gender-responsive indicators in policies from macro to micro level.
3. **Women are represented as vulnerable.** In most of the research on gender and climate change, women are defined as 'victims', 'vulnerable' or as 'environmental saviours'. The reason is the presumption of their dependency on natural resources due to their gender-specific roles (e.g. fetching water, cooking and securing wood for fuel).
4. **Integrating women but ignoring gender.** Women receive attention due to discrimination, 'being a woman'; however, men also suffer from discrimination and for gender studies, both men's and women's perspectives should be considered; though, at times, men's and women's capacities and strategies to address climate change are complementary (Lambrou and Nelson, 2010). Yet the programmes with mandatory gender inclusion are often inappropriate due to the following:
    a. **The male as a farmer.** Often, programmes take both men and women into account, though in reality they are based on a male perspective, believing males are farmers and women are caretakers of households. They believe policies and technologies should be designed in favour of men as they will be the ultimate users. In turn, the resultant increase in yield and income will benefit the female as a member of the household.
    b. **Standalone programmes for women's equality are all about women and projects focusing on women.** Often, research presents gender as women; supporting the comprehensive agenda to increase women's decision making, and hence power, has been a comprehensive agenda.
    c. **Gender (in number and intensity) is not focused.** Increasing women's involvement to participate in demonstrations and training is often used as a successful project indicator. Also, focus and presentation of results are often gender-binary (male and female). Simply ensuring numbers does not guarantee gender equality. Several guidelines are preferred for collection and documentation of gender-disaggregated data, yet adequate gender-disaggregated data remains a persistent problem. Even in gender-binary studies, women are often from female-headed households (FHH). Considering the position of women in developing countries, FHH in studies could refer either to *de jure* FHH (headed by divorced or widowed women) or *de facto*

FHH (in which the husband is absent temporarily, but may contribute to household finances). This suggests they are not comparable to men. Finally, there is a dearth of studies which focus on males alone or look outside the gender-binary concept:

> Gender equity should not be seen narrowly as a women's issue...it is an issue that requires men and women to work together in search of solutions that are both practical and based on principle. Increasingly, those solutions will be neither acceptable nor sustainable if the equal rights, dignity and worth of men and women are not respected.
>
> (Koïchiro Matsuura, Director-General of UNESCO, cited in Babugura, 2010)

There is still a common perception to many researchers that gender is the favouring of women's rights and responsibilities. In general, gender bias can be defined as a preference for or favouring one sex over the other (Rothchild, 2014). In trying to understand the reasons behind the insubordination of women in society, Ortner (1972) identifies three levels of the problem. The first refers to the universal fact of the inferior position of women in society, which is, however, socially and culturally endorsed. Second is the cultural symbolism, ideologies and social-structural arrangements that are related to women but differ from one culture to another. Thirdly, women revolt or try to suppress these structures of oppression and insubordination.

d. There is often a gender disconnect in project and programme cycles. Two contrasting thoughts for this issue exist. First, relatively strong gender analysis in the conceptual basis and planning of projects, on the one hand, and the much weaker integration of gender perspectives into implementation, monitoring and evaluation of the environment and climate change initiatives on the other (Otzelberger, 2011). Second, 'too often gender concerns are added into policies or processes as an afterthought and only focus on issues considered relevant to women' (Skinner, 2011).

5. Gender funding misdirection. Bryan *et al.* (2018) show that although policymakers are gradually recognizing the importance of gender in assessing climate vulnerability and developing effective CRS policies, lack of funding to support gender integration, lack of staff capacity on gender, and socio-cultural constraints act as barriers to the incorporation of gender into climate policies. The lack of an overarching structure or understanding of what these mechanisms and funds should be achieving often leads to inefficiencies and lack of co-ordination, making it difficult to ensure the implementation of gender policies. In 2018, for the first time, the Standing Committee on Finance (SCF) integrated gender substantively (Schalatek, 2020). The author outlined gender-responsive action for climate-financing instruments in climate agreements.

6. Masculinity. In most cases in which the nexus between gender and climate change *has* been investigated, the focus has been on women. There is almost no research on how climate change affects men. It is important to acknowledge that men also experience specific discrimination and vulnerability and they may face marginalization due to other aspects of their social identity such as their sexual identity and masculine value. These may make them more vulnerable than women in some cases (Box 3.3).

Box 3.3. Masculinity in strain: some examples.

'Traditionally, manliness is defined as a man's ability and responsibility to provide for his family and be financially secure and socially acclaimed' (Connell, 1995). Here we illustrate some of the examples in which indigenous masculinities may be both vulnerable and resilient in the face of climate change:

a) Rhoades (2003) points out that American Indian men are more likely to die from accidents than American Indian women in the event of severe weather or natural disasters, as the men are believed to perform heroic acts and are risk takers.
b) Scholars mention adoption of 'male migration' as a common coping strategy after disasters. However, their continued use and transmittance of their traditional knowledge (e.g. skills of catching fish) to the next generation are not considered in policies (Lee, 2003).
c) Eriksen *et al.* (2010), in their study in Australia, investigated the gendered dimensions of bushfires. They found that women were choosing not to take control of their own bushfire safety, whereas men were involved in fighting bushfires.
d) Another study in Australia, by Alston (2010), found that male farmers have higher suicide rates than the national average due to psychological distress of failed crops as a result of intense impacts of climate change on rural farms.
e) Gonda (2017) presents the case of Nicaragua where the common norm of being a 'real man' means being a cattle rancher. In 2013, government projects encouraged cattle ranchers to switch to cocoa production as a means of adapting to shorter rainy seasons. This intention was resisted by men, not because they are not concerned about the disastrous environmental outcomes and not because cocoa farming does not have long-term economic benefit, but because it took away symbolic masculine power from them.

7. One-size-fits-all policies for different contexts and gender identities. While there are global and national views on gender relations as defined in the gender wheel in Chapter 2, the relations are context-specific and change over time. Also, in a given setting, neither women nor men are a homogenous group; their relationship and roles change based on their capabilities and social variables such as age, marital status, income, etc. Policies often are made as 'one-size-fits-all' in all contexts, and need to be adjusted according to local needs.
8. 'Gender expert': being a woman is not enough. Although women often have an insight into gender issues, being a woman does not qualify someone to be a 'gender expert'. Also involving men is crucial for long-term and sustainable success. As suggested by Sheryl Sandberg in her blogs,[6] involvement of both sexes will benefit everyone, though men should not be part of 'Lean In' because it is nice for women to be managers to ensure equalization of opportunities and engagement in policies, with their skills for collaborative decision making at the household level.

Summary

This chapter draws attention to the gender-climate and gender-justice considerations in policies and institutional and legislative frameworks. There is increasing evidence of gender responsiveness in policies and programmes as well as in the representation of women in the programme bodies. Gender mainstreaming promotes gender equality and can only come through breaking down gender barriers and women's empowerment. The chapter provides a critical assessment of gender mainstreaming using the gender continuum tool. The discussion has been undertaken to understand the complex layers of policy drafting and implementation from macro to micro level with failure to achieve significant changes in the presence of several misconceptions of gender concepts.

Notes

[1] Source: Mary Robinson Foundation for Climate Justice,'Climate Justice Baseline', July 2013, p. 10. Available at: http://www.mrfcj.org/media/pdf/ClimateJusticeBaseline.pdf (accessed 20 April 2020). This report presents information gathered on the history of climate justice and differing perspectives on and approaches to climate justice.

[2] This maize sheller was 'ergonomically evaluated with farm women for shelling individual de-husked maize cobs. It is made of mild steel sheet and octagonal in shape' (Singh *et al.*, 2009).

[3] https://en.unesco.org/news/just-30-world%E2%80%99s-researchers-are-women-whats-situation-your-country (accessed 28 April 2020).

[4] https://unfccc.int/sites/default/files/resource/cp2021_04E.pdf (accessed 18 November 2020).

[5] https://genderchampions.com/

[6] Source: https://edition.cnn.com/2013/04/17/business/sandberg-gender-equality/index.html (accessed 7 March 2022).

References

Aguilar, M., Granat, C. and Owren (2015) *Roots for the Future: The Landscape and Way Forward on Gender and Climate Change*. IUCN and GCCA.

Alston, M. (2010) Rural male suicide in Australia. *Social Science & Medicine* 74, 515–522. DOI: 10.1016/j.socscimed.2010.04.036.

Babugura, A. (2010) *Gender and Climate Change: South Africa Case Study*. Heinrich Böll Stiftung, Cape Town, South Africa.

Bee, B., Biermann, M. and Tschakert, P. (2012) Gender, development, and rights-based approaches: lessons for climate change adaptation and adaptive social protection. In: Alston, M. and Whittenbury, K. (eds) *Research, Action, and Policy: Addressing the Gendered Impacts of Climate Change*. Springer, Dordrecht, The Netherlands, pp. 95–108.

Bryan, E., Bernier, Q., Espinal, M. and Ringler, C. (2018) Making climate change adaptation programmes in Sub-Saharan Africa more gender-responsive: insights from implementing organizations on the barriers and opportunities. *Climate and Development* 10(5), 417–431.

Connell, R.W. (1995) *Masculinities*. University of California Press, Los Angeles, California.

Dazé, A. and Dekens, J. (2017) A framework for gender-responsive National Adaptation Plan (NAP) processes (NAP Global Network Briefing Note). International Institute for Sustainable Development, Winnipeg, Canada. Available at: https://napglobalnetwork.org/wp-content/uploads/2017/07/napgn-en-2017-a-framework-for-gender-responsive-nap-processes.pdf (accessed 7 March 2022).

Eriksen, C., Gill, N. and Head, L. (2010) The gendered dimensions of bushfire in changing rural landscapes in Australia. *Journal of Rural Studies* 26, 332–342. DOI: 10.1016/j.jrurstud.2010.06.001.

FAO (2013) Project to Assist ERRA and its Partners in Restoring Livelihoods in Earthquake Affected Areas of Pakistan. Food and Agriculture Organization of the United Nations, Rome.

Available at: http://www.fao.org/emergencies/fao-in-action/projects/detail/en/c/180313/ (accessed 17 March 2022).

FAO and World Bank (2017) *How to integrate gender issues in climate-smart projects*. Training Module, FAO/World Bank, Rome.

Gay-Antaki, M. (2020) Feminist geographies of climate change: negotiating gender at climate talks. *Geoforum* 115, 1–10. DOI: 10.1016/j.geoforum.2020.06.012.

Gay-Antaki, M. and Liverman, D. (2018) Climate for women in climate science: women scientists and the Intergovernmental Panel on Climate Change. *PNAS* 115(9), 2060–2065.

Gonda, N. (2017) Rural masculinities in tension: barriers to climate change adaptation in Nicaragua. In: MacGregor, S. and Seymour, N. (eds) *Men and Nature: Hegemonic Masculinities and Environmental Change. RCC Perspectives: Transformations in Environment and Society* 4, 69–76.

Hemmati, M. and Röhr, U. (2009) Engendering the climate-change negotiations: experiences, challenges, and steps forward. *Gender & Development* 17(1), 19–32.

Huyer, S. (2016) *Gender and international climate policy: an analysis of progress in gender equality at COP21*. CGIAR Research Program on Climate Change, Agriculture and Food Security, Copenhagen.

Huyer, S., Twyman, J., Koningstein, M., Ashby, J., Vermeulen, S.J. *et al.* (2015) *Supporting Women Farmers in a Changing Climate: Five Policy Lessons*. CCAFS Policy Brief 10. Copenhagen. CGIAR Research Program on Climate Change, Agriculture and Food Security (CCAFS).

IGWG and USAID (2017) Gender integration continuum. Available at: https://www.igwg.org/wp-content/uploads/2017/05/FG_GendrIntegrContinuum.pdf (accessed February 2019).

IUCN (2013) The Environment and Gender Index (EGI) 2013 Pilot. IUCN Global Gender Office, Washington, DC. Available at: http://genderandenvironment.org/wp-content/uploads/2014/12/The-Environment-and-Gender-Index-2013-Pilot.pdf

IUCN and USAID (2016) Gender in mitigation action. EGI brief. IUCN Global Gender Office, Washington, DC. Available at: https://portals.iucn.org/union/sites/union/files/doc/gender_in_mitigation_actions4.pdf

Jackson, C. (1993) Doing what comes naturally? Women and environment in development. *World Development* 21(12), 1947–1963.

Kabeer, N. (2003) *Gender Mainstreaming in Poverty Eradication and the Millennium Development Goals: A Handbook for Policy-makers and Other Stakeholders*. Commonwealth Secretariat, London.

Lambrou, Y. and Nelson, S. (2010) *Farmers in a Changing Climate: Does Gender Matter?* Food and Agricultural Organization of the United Nations, Rome.

Lambrou, Y. and Piana, G. (2006) *Gender: The Missing Component of the Response to Climate Change*. Food and Agriculture Organization of the United Nations, Rome.

Leach, M. (2007) Earth mother myths and other ecofeminist fables: how a strategic notion rose and fell. *Development and Change* 38(1), 67–85. DOI: 10.1111/j.1467-7660.2007.00403.x.

Lee, M. (2003) 'How will I sew my baskets?' Women vendors, market art, and incipient political activism in Anchorage, Alaska. *American Indian Quarterly* 27(3/4), 583–592.

Mavisakalyan, A. and Tarverdi, Y. (2019) Gender and climate change: Do female parliamentarians make a difference? *European Journal of Political Economy* 26(1), 151–164.

NDC (2019) NDC Partnership Gender Strategy. Available at: https://ndcpartnership.org/sites/default/files/NDC%20Partnership%20Gender%20Strategy%202019.pdf (accessed March 2021).

Nelson, V., Meadows, K., Cannon, T., Morton, J. and Martin, A. (2002) Uncertain predictions, invisible impacts, and the need to mainstream gender in climate change adaptations. *Gender & Development* 10, 51–59.

O'Neill, S.J., Hulme, M., Turnpenny, J. and Screen, J.A. (2010) Disciplines, geography, and gender in the framing of climate change. *Bulletin of the American Meteorological Society* 91(8), 997–1002.

Ortner, S.B. (1972) Is female to male as nature is to culture? *Feminist Studies* 1(2), 5–31.

Otzelberger, A. (2011) *Gender-responsive strategies on climate change: recent progress and ways forward for donors*. BRIDGE, Institute of Development Studies, University of Sussex, Brighton, UK.

Pohlhaus, G. (2017) Varieties of epistemic injustice. In: Kidd, I.J., Medina, J. and Pohlhaus, G. (eds) *The Routledge Handbook of Epistemic Injustice*. Routledge, London, pp. 13–26.

Prebble, M., Gilligan, M. and Clabots, B. (2015) *Women's Participation in Global Environmental Decision Making: An EGI Supplemental Report*. IUCN, Washington, DC.

Rhoades, E.R. (2003) The health status of American Indian and Alaska native males. *American Journal of Public Health* 93(5), 774–778.

Rothchild, J. (2014) 'Gender bias'. In: Ritzer, G. (ed.) *The Blackwell Encyclopedia of Sociology*. Wiley-Blackwell, Oxford, UK.

Ruiz, F.J. and Vallejo, J.P. (2019) The post-political link between gender and climate change: The case of the nationally determined contributions support programme. *Contexto International* 41(2), 327–344.

Schalatek, L. (2020) Gender and Climate Finance. Climate Finance Fundamentals Heinrich Böll Stiftung, Washington, DC. Available at: https://climatefundsupdate.org/wp-content/uploads/2021/03/CFF10-ENG-2020-Digital.pdf (accessed October 2021).

SERVIR-Mekong (2015) *Gender and GIS: Guidance Notes*. Asian Disaster Preparedness Center, Bangkok.

Singh, S.P., Singh, R.S. and Agarwal, N. (2009) Women friendly improved farm tools and implements for commercialization. *Agricultural Engineering Today* 33(2).

Skinner, E. (2011) *Gender and Climate Change. Overview Report*. Institute of Development Studies, Brighton, UK.

UN (1992) Agenda 21. United Nations Conference on Environment & Development, Rio de Janeiro, Brazil, 3–14 June. Available at: https://sustainabledevelopment.un.org/content/documents/Agenda21.pdf (accessed 7 March 2022).

UNDP (2007) *Gender Mainstreaming: A Key Driver of Development in Energy and Environment Training Manual*. Energy and Environment Group, United Nations Development Programme, New York.

UNFCCC (2015) Decision-CP.20: Lima Work Programme on Gender. Available at: https://unfccc.int/files/meetings/lima_dec_2014/decisions/application/pdf/auv_cop20_gender.pdf (accessed May 2021).

UNFCCC (2016) Decision-CP.22: Gender and climate change. Available at: https://unfccc.int/files/gender_and_climate_change/application/pdf/auv_cop22_i15_gender_and_climate_change_rev.pdf (accessed May 2021).

Waters-Bayer, A., Kristjanson, P., Wettasinha, C., van Veldhuizen, L., Quiroga, G. *et al.* (2015) Exploring the impact of farmer-led research supported by civil society organisations. *Agricultural Food Security* 4, 4.

Williams, M. (2016) *Gender and Climate Change Financing: Coming Out of the Margin*. Routledge, London.

4 Farmers' Cognitive Perspectives in Dealing with Climate Risk: A Narrative Review

Abstract

Gender-responsive approaches are important to increase adoption (rate, behaviour and adaptive capacity) of climate-resilient strategies (CRS) to address climate change. Nonetheless, the psychological behaviour of individuals influences behavioural intention to adopt. This study uses new methodological approaches that go beyond gender being considered as biological and visible difference; for example, differences in terms of access and control over resources reduce individual capacity. The text includes theoretical and methodological approaches which consider individuals' behaviour and knowledge. The Protection Motivation Theory (PMT) approach is used to understand the cognitive behaviour of the farmer. Before adoption of CRS can take place, the farmer must accept and perceive the changes in, and components of, knowledge, experience and perception, which are a precursor to the adoption of CRS. Due to varying degrees of control as well as responsibilities, women and men have different adaptative capacities and other enabler (/barrier) factors which motivate (/discourage) them from the adoption (/dis-adoption) of any solutions.

4.1 Introduction

What drives farmers to make climate-relevant decisions to adopt or not adopt locally available and known climate-resilient technologies and policies? Despite the scientific consensus and farmers' acceptance of changes in weather and their dramatic impact on production and livelihood of farmers, differences among farmers are found in opting for potential and available climatic solutions. *Ceteris paribus*, the factors defining their identity (gender, age, social class), capabilities (economic status to buy the technology, farm size, etc.) and cognitive perspective influence their decisions. Their perspective makes them understand (/misunderstand) and interpret (/misinterpret) the climate risks and their capacity to deal with them. The two disciplinary dimensions that offer guidance for a cognitive perspective on CRS and adaptive capacity are: psychology (based on knowledge and

DOI: 10.1079/9781789249910.0004

experience) and behavioural intention. Fig. 4.1 presents the conceptual approach to understanding an individual's cognitive perspective. This chapter discusses the individual's psychological dimension of the cognition perspective for CRS and adaptive capacity using PMT. PMT constructs two pathways linking perceptions to behaviour; first, the 'threat appraisal pathway', in which a person assesses a threat's probability; and second, 'coping appraisal', in which a person evaluates their ability and self-efficacy to opt a CRS strategy to cope with the threat and cost associated with the strategy. This theory, led by Rogers, is one of the major theories within the domain of psychological research on health behaviour (Rogers, 1983; Rogers and Prentice-Dunn, 1997). The theory appears to have broad applicability, including natural and technological hazards.

As depicted in the above figure, *threat appraisal* has three components: knowledge, experience and perception. *Experience* is based on their practical exposure to the event which directly relates to them, or in some cases gained from looking at the experience of people in their surroundings. *Knowledge*, on the other hand, concerns the way people understand the world and how they interpret and apply meaning to their experiences (Blaikie *et al.*, 1997). The authors also mention that knowledge within a local population is not homogeneous but may vary from individual to individual depending on their gender, among other factors. The experiences are accumulated over time. The last component of threat appraisal is *perception*,[1] which has two sub-components, as discussed by Grothmann and Patt (2005): first, *perceived susceptibility*, i.e. the probability of occurrence of climate risk as well as whether the risk will affect them directly or indirectly; and, second, *perceived severity* is a person's appraisal of the extent of impact, i.e. how harmful the effects of climate risk would be on their farms or lives.

Like *threat appraisal*, *coping appraisal* also has three sub-components: adaptation efficacy, self-efficacy, and adaptation cost (Grothmann and Patt, 2005). Adaptation efficacy is routed in the belief that the use of adaptative strategies will reduce the impact of climate risk. Self-efficacy refers to the person's actual attitude, intentions and ability to decide and then carry out the adaptation strategies. Adaptation cost is the expected practical cost related to adaptation strategies and varies depending on their resource capacity and social capital. Government programmes for building adaptive capacity, especially in climate shocks (drought, flood or cyclones impacting areas and people), are a potential example of an enabler for technology adaptation.

Both threat and coping appraisal pathways guide behaviour and, consequently, farmers' decision making and actions in response to climate change. The decision of action or no action has two responses: adaptation and maladaptation. In the case of adaptation, the individual must deal with factors related to enabling, or placing barriers to, adaptation. Examples of barriers are unfamiliarity of the farmer with an available solution or lack of education, knowledge or the required skills to apply it.

Maladaptation, as defined in the IPCC Fifth Assessment Report, Working Group II (Impacts, Vulnerability, and Adaptation), is: '...*actions that may lead to increased risk of adverse climate-related outcomes, increased vulnerability to climate change, or diminished welfare, now or in the future*' (IPCC, 2014).

In the case of a farmer's decision to take no action and keep 'farming as usual', the maladaptation could be viewed as a response such as denial of the

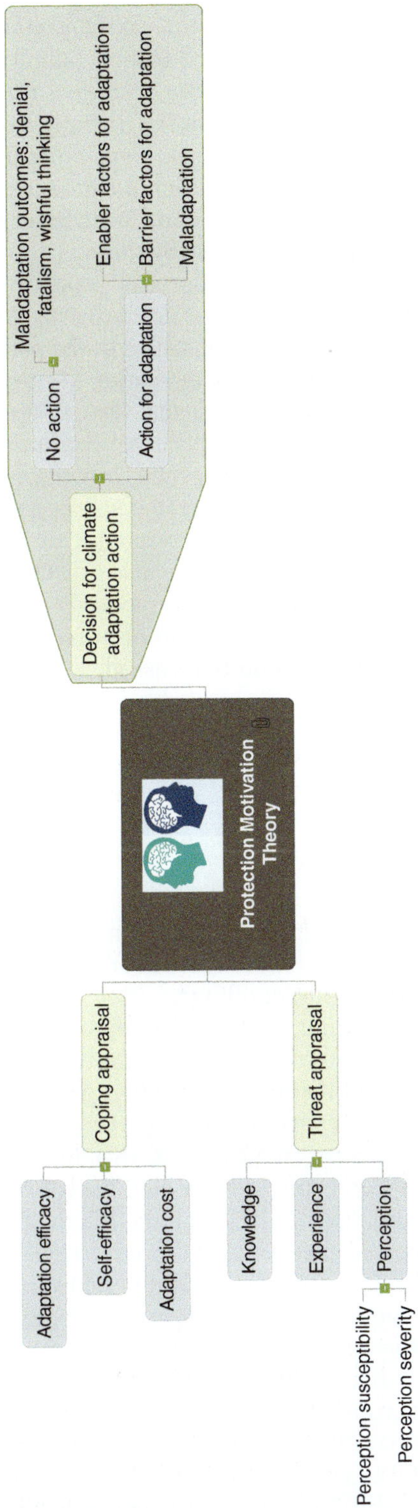

Fig. 4.1. Cognitive model for understanding an individual's decision in the context of climate risk. (From: the author, prepared using Mindview software)

threat, wishful thinking or fatalism. Those making maladaptive responses are usually resource-poor. In contrast, in the case of taking action, the maladaptation could be of many types as described by Neset *et al.* (2018).

The next part of the chapter builds on the narrative review[2] and provides evidence for each component discussed in the conceptual framework with gender-disaggregated responses. Because of gendered differences, men and women are found to differ in perceptions, experience and knowledge as well as differences in plausible responses preferred and adopted by them.

Most studies examined the 'threat appraisal' aspect by simply asking whether there is an increase or decrease, change or no change in climate attributes such as temperatures, precipitation, rainfall and others in the past few years. For perception severity, the questions asked are scale-based. Most of the time, self-efficacy questions in 'coping appraisal' are also scale-based. 'Self-efficacy is measured by directly probing the capability to perform a behaviour or indirectly based on beliefs about the ability to deal with specific inhibiting or facilitating factors' (Ajzen, 2002, p. 668). The other aspect, depending on the research objective, could be examined with binary (yes/no) responses or open-ended questions or with a selection of answers from a pre-response list. The responses are based on who answers the question and his or her responses may vary according to their own experiences, knowledge, character and resource allocation. Table 4.1 illustrates the potential set of generic questions asked.

4.2 Protection Motivation Theory

4.2.1 Threat appraisal

4.2.1.1 Experience

The majority of the studies exploring *experience* are based on their practical exposure to the event which directly relates to them or, in some cases, on looking at the experience of people in their surroundings.

Twyman *et al.* (2014) found, in all four surveyed sites in Africa, that both adult male and female decision makers mentioned that they have observed changes in climate change during their lifetime, though there are significant and contrasting statistical differences in the types of events experienced by them across the sites in the last five years. For example, in Nyando, Kenya, more women have reportedly experienced more floods and storms than men have, whereas men report more drought and erratic rainfall. In Rakai, Uganda, the majority of both men (70%) and women (87%) reported experiencing drought. This difference may be routed back to the gender wheel and the distinctly different work that men and women do, largely dictated by gender norms. They perceive changes differently. For example, as mentioned by the authors, women may have mentioned: 'more droughts or less rainfall because they walk farther to collect water and have less water for producing subsistence crops'. A similar argument about experience and reason is given by Kanda *et al.* (2017), where they mentioned more drought or dryness in the summer season, and as a result women must cover more distance. The authors found 63% of females and 73% of males mentioned that water collection is a female activity and no males collect it. Similarly, Mehar (2014) found significant

Table 4.1. Exemplary cognitive questions.

Variable	Potential survey questions
Knowledge	Have you heard about climate change? What do you think are the reasons behind climate variation?
Experience	Have you noticed any changes in climate or weather patterns in the last *X* years? If yes, what changes have you noticed over your lifetime? Did the [event name] occur in last season? How many times did the [event name] occur in the last *X* years? Or: For how many years has [event name] been a problem? Did (or how did) the [event name] affect your agricultural activities? Did the [event name] result in a significant reduction in food consumption or income levels of your household?
Perceived susceptibility	Do you believe climate change could affect your agricultural activities/crop production/livelihood if you do not prepare? To what extent do you agree with the statement that [event name] is a problem for your plot? (Options: strongly agree/agree/neutral/disagree/ strongly disagree)
Perceived severity	What possible impacts do you anticipate from these changes? Do you believe climate change can significantly endanger your life? Do you anticipate climate change will result in crop loss and no food for consumption? What do you expect the impacts, if any, of such future changes to be?
Adaptation efficacy	Do you have enough resources to cope with climate shock (after the event)? Do you have enough resources to use for climate adaptation strategies (pre-event)? Do you or your family members have a plan for what to do to protect yourselves in the event of extreme impacts of climate change (pre-event)? Does your household have emergency kits that include items such as stored grain, drinking water and other essential things that can sustain you for a few days in the case of extreme events associated with climate change (pre-event)?
Self-efficacy	Do you think that you can protect yourself from dangerous events caused by climate change?
The decision for action	Are you planning to take any action to reduce climate risk? Have you planned to make any changes in agricultural activities because of any previous climate change-related impacts? What changes have you planned or opted for in the last season? (Options: crop rotation, change in sowing/harvesting time, change in crop or variety, etc.)
Enablers of or barriers to adoption of CRS	Are there serious obstacles and barriers to hinder you from protecting yourself from the negative consequences of climate change? Is there any supporting infrastructure or enabler to help you protect yourself from climate risk? Do you think you have the necessary information and resources to prepare for climate change? If yes, what are these?

gender differences while reporting changes in weather in the last 15–20 years in Bihar, India. Both male and female farmers reported high temperature followed by less rainfall, heatwaves, frequent droughts and that hot months are getting hotter. The gender difference for all other effects was not significant, but more men reported high temperature, hot months getting hotter and less rainfall. In another study, in Bangladesh, Rakib *et al.* (2017) found, 'among the rural communities, children, women and disabled people have more experience of flood-related hazards and the risks associated with it'.

The experience is not limited to climate change *per se* but also to the direct and indirect impacts they face. For example, Salvati *et al.* (2018) found landslides and floods killed more males than females, though there is variation across the age group (Fig. 4.2). Studies (Neumayer and Plümper, 2007; Wright *et al.*, 2015) suggest that women tend to die younger than men because of the consequences of climate change. On the other hand, men are affected differently. Other studies have also shown the strong links between climate-related disasters and female mortality, with women, boys and girls more than 14 times more likely than men to die during a disaster (Peterson, 2007; Bradshaw and Fordham, 2013). The UN estimates that women account for approximately 80% of those who are displaced by climate change.

It is important to note that 'individuals experience more emotive mental imagery concerning short-term goals that are more engaging than long-term

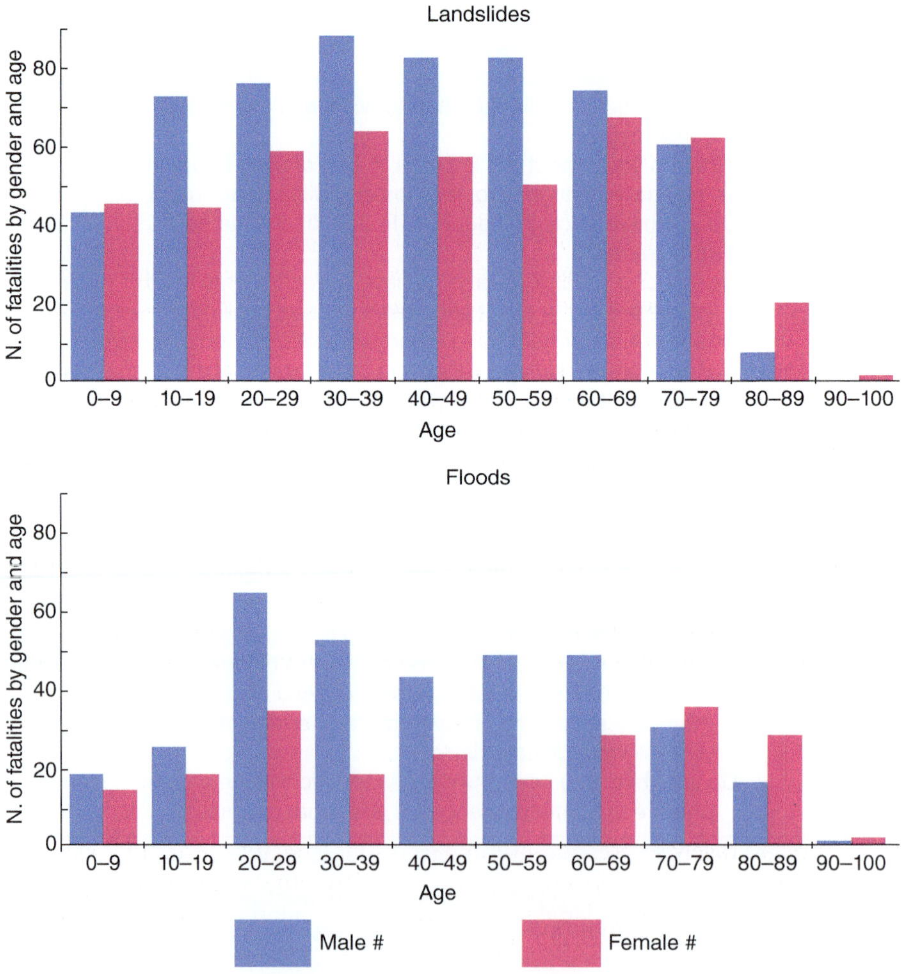

Fig. 4.2. Fatalities by gender and age in the 50 years 1965–2014 in Italy (%). (Adapted with permission from Salvati *et al.*, 2018)

goals that are less specific' (Pahl *et al.*, 2015), and this may be a reason why few studies argue differences between actual climate data and farmer-reported data.

4.2.1.2 Knowledge

Based on their own farming experience and observed impact on crops and livelihoods in past climate events, farmers have varying sense and knowledge, and varying degrees of accuracy, about climate change occurrence, its impact, causes and consequences. Different crops, with different stages of crop development, and livestock species have their varying sensitivities to climate change (type and intensity). This knowledge, coupled with farmers' experience of climate change, helps them decide which species to grow. For example, Paris *et al.* (2008) suggest that ignoring women's indigenous knowledge and preferences for rice varieties may lead to slow adoption of new varieties. Men and women farmers have distinct and critical knowledge to contribute to disaster management (FAO).[3] Another set of knowledge is about the skills of farming or coping activities. For example, in a few places native to tsunami-affected countries in 2004, girls and women were generally not being taught key survival skills such as swimming and climbing trees, and the gendered practices of women keeping long hair and wearing saris have an adverse effect during climate shocks, like the above-mentioned tsunami (Murty *et al.*, 2006).

4.2.1.3 Perception

Women and men perceive and assess risks differently, and that is also true for climate change. The accuracy of perception depends on knowledge and experience. Their perception depends on their experience and knowledge of identifying natural changes like direction and speed of wind; cloud formation, density and movement of clouds, etc.

Perceived susceptibility studies have overall similar results to how farmers responded about the experience, especially for temperature and precipitation. For instance, farmers in southern Africa and Bangladesh perceived that the regions were getting warmer and drier, with increased frequency of droughts and changes in timings of rains, in the last couple of years (Rakib *et al.*, 2017). The authors have also reported that females' perception could be more accurate than males'. Further, findings by Ishaya and Abaje (2008) on perceptions of climate change in Nigeria revealed that indigenous people perceived that the climate had been changing over the years due to diverse human activities. Ngigi *et al.* (2017) asked about the perceptions of both husbands and wives. They found that both perceived a decrease in rainfall. Wives, in contrast, perceived an increased incidence of erratic rainfall with profound flooding and early onset of rain.

Studies exploring perceived severity contained scale-based responses. For example, Adzawla *et al.* (2019) asked farmers in Ghana: 'In your opinion what is the level of severity of climate change on farming activities in your communities?' (options: very severe, averagely severe, not severe). The results were that, 'except for female heads (who reported very severe impact), over half of the other gender groups revealed that climate change had mild impacts on farming'. The reasons may again be due to different roles of women and men including *ex-post* climate change. For example, in the post-flood recovery stage, women in Bangladesh contribute by homestead gardening and livestock rearing, followed household repairing and taking care of health and livestock diseases, while male participation was observed to be not as substantial.

Farmers' perceptions are considered subjective, negating the role of perception in adaptive decision making, which may have a counterproductive maladaptation effect. This may result in a vicious circle of social inequalities whereby women are viewed as victims of climate change. More women affected by climate change and more inequality mean more adverse impact of climate change.

4.2.2 Coping appraisal

As mentioned by Grothmann and Patt (2005), coping appraisal, or perceived adaptive capacity, 'comes after the risk perception process and only starts if a specific threshold of threat appraisal is passed'. 'A minimum level of threat or concern must exist before people start contemplating the benefits of possible actions and ruminate their competence to perform them' (Schwarzer, 1992, p. 235, quoted in Grothmann and Patt, 2005)

4.2.2.1 Adaptive efficacy

Adaptive efficacy determined by adaptive capacities of individuals greatly depends on their income, education, health and access to natural resources. Given that women tend to be poorer, less educated, have a lower health status and have limited direct access to or ownership of natural resources, they will be less likely to adopt a new practice, especially when they have to buy it. First, the overall attitude towards adaptation is low. Intuitively, the majority of studies report male farmers, compared to female, are more responsive to adapting to climate change. They not only adopt technology but they also have a different attitude towards it. But all men and women are not alike in their experience of and responses to climate change (Terry, 2009; Bee, 2016; Mersha, 2018). For instance, in their study in northern Mali, Djoudi and Brockhaus (2011) identified that not only do men and women use different CRS strategies, but they also have a different perception of the same strategies. According to their findings, rural men consider migration as a CRS action, whereas women consider it the cause of vulnerability. Box 4.1 elaborates on vulnerability and gender inequality.

Moreover, evidence indicates that men and women encounter different adaptation constraints. For instance, in their study in Uganda, Ghana and Bangladesh, Jost *et al.* (2016) find that women have less adaptive capacity than men, as they encounter resource constraints, limited access to information and extension services, and high labour demands in available CRS choices. Bee (2016) also demonstrates the differentiated effect of 'changing institutions, politics and landscapes' on women in a study carried out in rural Mexico.

Few studies demonstrate that women often raise crops that are more sensitive to climate variability than do men, and use their indigenous knowledge for farming and climate solutions.

4.2.2.2 Self-efficacy

Adaptation capacity is not sufficient to motivate people to adopt; their belief that they can do specific tasks is a crucial condition. This perceived self-efficacy reflects belief in their capability and social environment. However, it should not be confused with self-esteem and self-confidence. Self-esteem refers to general feelings of

Box 4.1. Vulnerability and gender inequality.

The IPCC Third Assessment Report (TAR) describes vulnerability as:

> The degree to which a system is susceptible to, or unable to cope with, adverse effects of climate change, including climate variability and extremes. Vulnerability is a function of the character, magnitude, and rate of climate variation to which a system is exposed, its sensitivity, and its adaptive capacity (IPCC, 2001, p. 995).

The three dimensions of vulnerability to climate change are:

- ***exposure*** – the degree to which people and the things they value could be exposed to climate variation or change;
- ***sensitivity*** – the degree to which they could be harmed by that exposure; and
- ***adaptive capacity*** – the degree to which they could mitigate the potential for harm by taking action to reduce exposure or sensitivity.

How households perceive a climate shock and react to it is an important lens through which to assess vulnerability and its potential impact on the household. In gender-climate studies, women are mostly portrayed as victims or vulnerable in relation to its impacts. The IPCC also documents impact studies and predicts that 'climate change will also exacerbate existing gender inequalities, adding to the vulnerabilities already experienced by rural women in the developing world'. This victimization discourse emphasizes the vicious circle of the feminization of vulnerability, i.e. climate change negatively affects women more than men due to pre-existing gender inequality – the unequal impact reinforces the gender and social inequality. This creates a further imbalance in resources and power, which magnifies the impact. When this is true, that women are impacted most, a vicious circle may occur; portraying women as vulnerable may misrepresent the 'causes of vulnerability and obscure the role of women as proactive agents of climate change adaptation' (Alston, 2020). This is, 'in our discussions of vulnerability in the climate change context, to avoid assigning passivity and victimhood to women and address agency and emancipation' (ibid.). Djoudi *et al.* (2016) suggest that women dominate discussion of vulnerability which ignores its multi-layered complexity. The multi-layered complexities mediated by intersectional factors (such as class, age, wealth, activities involvement, marital status, poverty level, religion, power relations, etc.) are important to consider.

self-worth or self-value (Kling *et al.*, 1999). Self-confidence is a person's belief in his/her own worth and it is a combination of self-esteem and general self-efficacy (Mehar and McDougall, 2018).

Schmuck-Widmann (1996) mentioned that during a cyclone warning, women alone at home do not think that they can leave the house without their husbands, not only due to religious norms but also due to their normative beliefs of moving alone. In the recent past, various initiatives have encouraged women's participation and empowerment to raise their voices and exercise leadership within their communities for climate-disaster management practices. Women often have low levels of confidence, shyness and hesitation to speak, all of which prevent them from participating in public meetings (Sogani, 2016, p. 50). Climate adaptation initiatives have found increasing levels of confidence in women, a factor that has contributed to the sustainability of intervention. In the Women's Empowerment: Improving Resilience, Income and Food Security (WE-RISE) programme in Ethiopia, there was a 91% increase in women's self-confidence and a 56% increase in their access to decision making within the household (FAO and CARE, 2019).

4.2.2.3 Adaptation cost

New research highlights the fact that technologies and practices adopted at the household level do not benefit all members of the household equally (Theis *et al.*, 2017). Some practices, such as conservation agriculture, may have a negative impact on women due to increased labour requirements (Nelson and Stathers, 2009; Beuchelt and Badstue, 2013). Climate change responses can affect women both positively and negatively, suggesting that there are important trade-offs across outcomes that must be considered. For example, male out-migration as an adaptive response to climate change may increase women's decision-making authority while at the same time increasing their labour burden (Nelson and Stathers, 2009; Djoudi and Brockhaus, 2011).

The other dimension of adaptation cost depends on the barrier and enabler factors associated with the research context as well as technologies.

4.3 The Decision for Climate Adaptation

Both threat and coping appraisal pathways act as a precursor to adoption of CRS. Farmers deciding not to take any action to adopt mitigation strategies for future climate change may have counterproductive results; for example, having no food due to crop loss. Another example is barriers to girls' education, as they have to stay home to carry heavy loads of water and other things such as fodder, wood for fuel, etc. for long distances, perpetuating the cycle of women's disempowerment. In some of these cases, there is increasing evidence of sexual harassment and assault faced by women from, for instance, forest owners, managers or guards. Obonyo and Mogoi (2009) found that 42% of women mentioned harassment by forest guards and also rape.

The reasons for adoption varied differently in different countries and by gender, a fact attributable to the perceived levels of severity of each constraint and the expected benefits, which are likely to be influenced by socio-economic characteristics and, in turn, influence farmers' perceptions (Prager and Posthumus, 2010). According to Hassan and Nhemachena (2008), women are more responsive in adapting to climate change. One possible reason for this observation is that in most rural smallholder farming communities in the region, men are more often based in towns, and much of the agricultural work is done by women. The adoption of CRS choices is explored in more detail, with supporting evidence, in Chapter 7.

The literature indicates that, along with the above reasons, there are also variations in barrier and enabler factors motivating farmers to take a decision. The enabler factor could be supporting the environment, access to market, kinship networks or network-based information. Again, social norms like mobility constraints and not speaking to another adult male outside the family restrict women from knowledge and skills related to adaptation strategies. The other barrier factors with varying degrees for both men and women range from lack of formal credit or insurance access to informal credit sources. Fletschner's (2009) study in rural Paraguay found that 23% of women declared that they were credit-constrained compared to 17% of men. The barrier and enabler factors, as well as examples of maladaptation after deciding on climate coping strategies, are discussed in Part IV of this book.

Barrier and enabler factors vary across individuals, cultures and economic backgrounds. This requires removing the generalizations or assumptions which generalize the experiences of individuals to all women and men. This emphasizes the need to consider gender analysis with an intersectionality lens (Box 4.2). Above all, and mostly

Box 4.2. The new wave of intersectionality and gender.

The intersectional theory was introduced in the late 1980s by Crenshaw (1991) who described how 'race and gender interact in contexts of violence against women of colour'. These sentiments influence many scholars to see the intersection of gender with many other factors such as age, class, ethnicity, access to resources, legal status on land and other social divisions. According to McCall (2005), intersectionality rejects the single-axis framework that investigates gender and recognizes the complexity of social life (complex inequalities) and captures the 'the relationship among multiple dimensions and modalities of social relations and subject formations'.

Other formal definitions of intersectionality are: 'The influences of multiple identities in a person as these interact with marginalizing and empowering structures, norms and narratives' (Colfer *et al.*, 2018, p. 2); and: 'It conceptualizes social categories as interacting with and co-constituting one another to create unique social locations that vary according to time and place. These intersections and their effects are what matter in an intersectional analysis' (Hankivsky, 2014, p. 9).

Indeed, intersectionality has provided a framework to describe how cultural, social and political forces influence women's and men's different exposure to the impact of climate and responses and capacities to cope with them. For example, Nation (2010) narrates women's lack of proper rights of access to land and water which limits them in participating in adaptation strategies. In another example, by Moula (2012), higher education increases the probability of adopting resilient strategies for both female- and male-headed households. The author argues that education increases their ability to understand the information and to make an innovative decision. Thus, intersectionality is an important aspect in deciding CRS choices for climate change in agrarian settings.

missed in studies, is farmers' concerns about climate variability within crop stages rather than inter-crop seasonal variation. This is explicitly discussed in Chapter 9.

4.4 Summary

The negative effects of climate change are realized in varying degrees by farmers globally. It is argued that in addition to limitations presented by the availability of technologies, people's perceptions and knowledge within a society can fundamentally limit climate change adaptation (Adger *et al.*, 2009). The answer to what drives people in making such decisions is rooted in the cognitive model as presented in this chapter. The chapter discusses the knowledge and behaviour of farmers from a broad perspective, though with examples referring to context. There is a dearth of literature for livestock and fisheries.

There are two criticisms often faced by different branches/components of the cognitive model. The first is self-reported data which may have bias due to personal experience and recall. Farmers usually report from their experience of recent events rather than their spread or effect over the whole year. Few studies suggest actual climate data are different from that stated by farmers, though they may have their limitations when analysing. Sometimes farmers also misinterpret terms; for example 'weather' and 'climate change' are used interchangeably. Some farmers, mostly influenced by their belief in God or fate, perceive the impact will be negligible or low for them. The second criticism is that men and women often under-report or manipulate responses in accordance with social norms and underlying fears – often, women who disobey their husbands or break social norms may suffer domestic violence. Men, on the other hand, who break social norms or help their family by contributing to the work normally done by the wife may be subject to social taboos or being laughed at by other men, and even other women, in the society.

Notes

[1] Perception refers to beliefs or opinions held by people based on how things seem to them (*Cambridge Advanced Learner's Online Dictionary and Thesaurus*, 2011).
[2] Narrative review, selective in nature, aims to summarize or synthesize what has been written on a topic but does not seek generalization or cumulative knowledge from what is reviewed (Paré *et al.*, 2015). Petticrew and Roberts (2006) defined this as 'the process of synthesizing primary studies and exploring heterogeneity descriptively, rather than statistically'.
[3] fao.org/docrep/fao/010/a1395e/a1395e00.pdf (accessed 11 March 2022)

References

Adger, W.N., Dessai, S., Goulden, M., Hulme, M., Lorenzoni, I. *et al.* (2009). Are there social limits to adaptation to climate change? *Climatic Change* 93, 335–354.
Adzawla, W., Azuman, S.B., Anani, P.Y. and Donkoh, S.A. (2019) Gender perspectives of climate change adaptation in two selected districts of Ghana. *Heliyon* 5(11), e02854.
Ajzen, I. (2002) Perceived behavioral control, self-efficacy, locus of control, and the theory of planned behavior. *Journal of Applied Social Psychology* 32, 665–683.
Alston, M. (2020) Gendered vulnerabilities and adaptation to climate change. In: Sacs, C., Sexsmith, K. and Jensen, L. (eds) *Gender and Agriculture Handbook*. Routledge, Abingdon, UK.
Bee, B. (2016) Power, perception, and adaptation: exploring gender and social–environmental risk perception in northern Guanajuato, Mexico. *Geoforum* 69, 71–80.
Beuchelt, T.D. and Badstue, L. (2013) Gender, nutrition- and climate-smart food production: opportunities and trade-offs. *Food Security* 5, 709–721.
Blaikie, P., Brown, K., Stocking, M., Tang, L., Dixon, P. and Sillitoe, P. (1997) Knowledge in action: local knowledge as a development resource and barriers to its incorporation in natural resource research and development. *Agricultural Systems* 55(2), 217–223.
Bradshaw, S. and Fordham, M. (2013) *Women, Girls and Disasters: A Review for DFID.* Department for International Development (DFID), London.
Colfer, C., Sijapati Basnett, B. and Ihalainen, M. (2018) *Making Sense of 'Intersectionality': A Manual for Lovers of People and Forests*. Occasional Paper 184. CIFOR, Bogor, Indonesia.
Crenshaw, K.W. (1991) Mapping the margins: intersectionality, identity politics, and violence against women of color. *Stanford Law Review* 43, 1241–1299.
Djoudi, H. and Brockhaus, M. (2011) Is adaptation to climate change gender neutral? Lessons from communities dependent on livestock and forests in northern Mali. *International Forestry Review* 13(2), 123–135.
Djoudi, H., Locatelli, B., Vaast, C., Asher, K., Brockhaus, M. and Basnett, B. (2016) Beyond dichotomies: gender and intersecting inequalities in climate change studies. *Ambio* 45, 248–262.
FAO and CARE (2019) *Good Practices for Integrating Gender Equality and Women's Empowerment in Climate-Smart Agriculture Programmes*. CARE, Atlanta, George.
Fletschner, D. (2009) Rural women's access to credit: market imperfections and intrahousehold dynamics. *World Development* 37(3), 618–631.

Grothmann, T. and Patt, A. (2005) Adaptive capacity and human cognition: the process of individual adaptation to climate change. *Global Environmental Change* 15, 199–213.

Hankivsky, O. (2014) *Intersectionality 101*. The Institute for Intersectionality Research and Policy, Simon Fraser University, Vancouver, Canada.

Hassan, R. and Nhemachena, C. (2008) Determinants of African farmers' strategies for adaptation to climate change: multinomial choice analysis. *African Journal of Agricultural and Resource Economics* 2(1), 83–104.

IPCC (2014) *Climate Change 2014: Synthesis Report. Contribution of Working Groups I, II and III to the Fifth Assessment Report of the Intergovernmental Panel on Climate Change*. (edited by Pachauri, R.K. and Meyer, L.A.). IPCC, Geneva, Switzerland, p. 151.

IPCC (2001) *Climate Change 2001: Impacts, Adaptation, and Vulnerability. IPCC Third Assessment Report. Contribution of Working Group II to the Third Assessment Report of the Intergovernmental Panel on Climate Change*. IPCC, Geneva, Switzerland. Available at: https://www.ipcc.ch/site/assets/uploads/2018/03/WGII_TAR_full_report-2.pdf (accessed 8 March 2022).

Ishaya, S. and Abaje, I.B. (2008) Indigenous people's perception on climate change and adaptation strategies in Jema'a local government area of Kaduna State, Nigeria. *Journal of Geography and Regional Planning* 1(18), 138–143.

Jost, C., Kyazze, F., Naab, J., Neelormi, S., Kinyangi, J., *et al.* (2016) Understanding gender dimensions of agriculture and climate change in smallholder farming communities. *Climate and Development* 8(2), 133–144.

Kanda, A., Murongazvombo, M. and Ncube, F. (2017) Adapting household water use in rural Zimbabwe. *International Journal of Environmental Studies* 74(3), 471–485.

Kling, K.C., Hyde, J.S., Showers, C.J. and Buswell, B.N. (1999) Gender differences in self-esteem: a meta-analysis. *Psychological Bulletin* 125, 470–500.

McCall, L. (2005) The complexity of intersectionality. *Signs Journal of Women in Culture and Society* 30(3), 1771–1800.

Mehar, M. (2014) Gender specific impacts of climate variation in agriculture. Unpublished MPhil thesis, Department of Economics, School of Social Science, Indira Gandhi National Open University, India.

Mehar, M. and McDougall, C. (2018) *Methods and Tools for Gender Analysis in FISH: A Preliminary Consolidation and Reference Guide*. Penang, Malaysia: CGIAR Research Program on Fish Agri-Food Systems.

Mersha, A.A. (2018) Unpacking the nexus between gender and climate change adaptation: the case of smallholder agriculture in north-eastern Ethiopia. PhD thesis, Utrecht University Repository. Available at: https://dspace.library.uu.nl/handle/1874/364787

Molua, E. (2012) Gendered response and risk-coping capacity to climate variability for sustained food security in Northern Cameroon. *International Journal of Climate Change Strategies and Management* 4, 277–307.

Murty, T.S., Aswathanarayana, U. and Nirupama, N. (2006) *The Indian Ocean Tsunami*. Taylor & Francis, Balema, The Netherlands.

Nation, M. (2010) Understanding women's participation in irrigated agriculture: a case study from Senegal. *Agriculture and Human Values* 27(2), 163–176.

Nelson, V. and Stathers, T. (2009) Resilience, power, culture, and climate: a case study from semi-arid Tanzania and new research directions. *Gender and Development* 17(1), 81–95.

Neset, T.S., Wiréhn, L., Klein, N., Käyhkö, J. and Juhola, S. (2018) Maladaptation in Nordic agriculture. *Climate Risk Management* 23, 78–87.

Neumayer, E. and Plümper, T. (2007) The gendered nature of natural disasters: the impact of catastrophic events on the gender gap in life expectancy, 1981–2002. *Annals of the Association of American Geographers* 97(3), 551–566.

Ngigi, M.W., Mueller, U. and Birner, R. (2017) Gender differences in climate change adaptation strategies and participation in group-based approaches: an intra-household analysis from rural Kenya. *Ecological Economics* 138, 99–108.

Obonyo, E. and Mogoi, J. (2009) Integrating gender in forest management in Kenya. Paper prepared for the World Agroforestry Congress, 18–23 October, Buenos Aires. Kenya Forestry Research Institute, Nairobi.

Pahl, S., Sheppard, S., Boomsma, C. and Groves, C. (2015) Perceptions of time in relation to climate change. *Wiley Interdisciplinary Reviews: Climate Change* 6(3), 375–388.

Paré, G., Trudel, M.C., Jaana, M. and Kitsiou, S. (2015) Synthesizing information systems knowledge: a typology of literature reviews. *Information & Management* 52(2), 183–199.

Paris, T.R., Singh, A., Cueno, A.D. and Singh, V.N. (2008) Assessing the impact of participatory research in rice breeding on women farmers: a case study in eastern Uttar Pradesh, India. *Experimental Agriculture* 44(1), 97–112.

Peterson, K. (2007) *Reaching Out to Women when Disaster Strikes*. Soroptimist White Paper, Soroptimist International, UK.

Petticrew, M. and Roberts, H. (2006) *Systematic Reviews in the Social Sciences: A Practical Guide*. Blackwell Publishing, Oxford, UK.

Prager, K. and Posthumus, H. (2010) Socio-economic factors influencing farmers' adoption of soil conservation practices in Europe. In: Napier, T.L. (ed.) *Human Dimensions of Soil and Water Conservation*. Nova Science Publishers, pp. 203–223.

Rakib, M.A., Islam, S., Nikolaos, I., Bodrud-Doza, Md. and Bhuiyan, A.H. (2017) Flood vulnerability, local perception and gender role judgement using multivariate analysis: a problem-based 'participatory action to Future Skill Management' to cope with flood impacts. *Weather and Climate Extremes* 18, 29–43.

Rogers, R.W. (1983) Cognitive and physiological processes in fear appeals and attitude change: a revised theory of protection motivation. In: Cacioppo, J.T. and Petty, R.E. (eds) *Social Psychophysiology: A Sourcebook*. Guilford Press, New York, pp. 153–177.

Rogers, R.W. and Prentice-Dunn, S. (1997) Protection motivation theory. In: Gochman, D. (ed.) *Handbook of Health Behavior Research*. Plenum Press, New York.

Salvati, P., Petrucci, O., Rossi, M., Bianchi, C., Pasqua, A.A. and Guzzetti, F. (2018) Gender, age and circumstances analysis of flood and landslide fatalities in Italy. *Science of the Total Environment* 610–611, 867–879.

Schmuck-Widmann, H. (1996) *Living with the Floods: Survival Strategies of Char-Dwellers in Bangladesh*. FDCL, Berlin.

Schwarzer, R. (1992) Self-efficacy in the adoption and maintenance of health behaviours: theoretical approaches and a new model. In: Schwarzer, R. (ed.) *Self-Efficacy: Thought Control of Action*. Hemisphere, Washington, DC, pp. 217–243.

Sogani, R. (2016) Gender approaches in climate compatible development: lessons from India. Available at: https://cdkn.org/wp-content/uploads/2016/05/CDKN_India-Gender_report_WEB.pdf (accessed 8 March 2022).

Terry, G. (2009) No climate justice without gender justice: an overview of the issues. *Gender & Development* 17(1), 5–18.

Theis, S., Lefore, N., Meinzen-Dick, R. and Bryn, E. (2017) *What Happens After Technology Adoption? Gendered Aspects of Small-Scale Irrigation Technologies in Ethiopia, Ghana, and Tanzania*. IFPRI Discussion Paper 01672. International Food Policy Research Institute, Washington, DC.

Twyman, J., Green, M., Bernier, Q., *et al.* (2014) *Gender and Climate Change Perceptions, Adaptation Strategies, and Information Needs Preliminary Results from four sites in Africa*. CCAFS Working Paper no. 83. CGIAR Research Program on Climate Change, Agriculture and Food Security (CCAFS), Copenhagen.

Wright, H., Huq, S. and Reeves, J. (2015) *Impact of Climate Change on Least Developed Countries: Are the SDGs Possible?* International Institute for Environment and Development (IIED). IIED Briefing, London. Available at: http://pubs.iied.org/pdfs/17298IIED.pdf (accessed 1 June 2020).

5 Gender in Dealing with Climate Risk: Data Triangulation Evidence

Abstract

Multiple data-collection techniques were used to increase the validity and reliability of the research through triangulation. This also benefits low-budget programmes. The present chapter uses two different data sources compiling different understandings about climate change events with a gender dimension.

5.1 Introduction

'Neither the impacts of climate change on people nor the ways in which people respond to climate change are gender-neutral' (Otzelberger, 2011). Numerous in-depth and country-specific studies pertaining to the above areas have been carried out, as reviewed in Chapter 3. Although, across studies, there is consistency in the approaches used and the questions in the context of different components of the cognitive model frame, as in the previous chapter, the majority of these studies are regional and portray things relevant at local or national level. Other than issues of bias in subjective and self-reported responses for perception, knowledge and experience, the studies are often characterized by a general paucity of data and a significant heterogeneity in the reliability and representativeness of data at a national or global level.

Given that adoption of any technology is influenced by perception, and, in turn, this is influenced by gender, knowledge of these relationships is a prerequisite to intensive upscaling and assimilation of descriptive information about individuals' cognitive behaviour and can lay the foundation for building policies and implementation strategies for climate-resilient strategies (CRS). In line with the above arguments, the present study presents the relevant evidence using the triangulation approach based on data available in the public domain.

 Climate Change, Adaptation and Gender: Policy, Practice and Methodological Underpinnings (M. Mehar and N. Prasad)
DOI: 10.1079/9781789249910.0005

5.2 Materials and Methods

5.2.1 Triangulation

Lately, scholars have advocated use of different sources of data within a single study with the aim to obtain a bigger and more complete picture of the topic under investigation. The process of combining data on a particular phenomenon is defined as triangulation. Triangulation typically refers to the integration of more than one method to add breadth and depth to data and their analysis, thus enhancing confidence in the findings (Bryman, 2009). Data gathered from multiple sources are not always used for confirmatory purposes and to increase validity. They are also used to secure a more complete picture, as using only one source might miss some important perspective on the investigated phenomenon. There is no inherent limit to the size of different data compiled.

Ideally, triangulation is useful when immediate knowledge about the phenomenon is needed and the collection of new data is not feasible or cost-effective. There are four types of triangulation: we use data triangulation here (Box 5.1). Data triangulation uses multiple sources of data; each source may have its own research method and may represent different facets of the phenomenon.

5.2.2 Data sources

The triangulation approach illustrated here is compiling information from two different secondary data sources from India. These data sources are available in the public domain. All the data sources used in this chapter are largely cross-section primary household surveys. In all the data sets, both male and female farmers were interviewed, though between these two data sets and within a data set, multiple gender identities within a household were studied. The studies

Box 5.1. Triangulation types.

Four basic types of triangulations defined are (Denzin, 1989, 2006):

1. data triangulation: involves using multiple sources of data (time, space and persons) in an investigation;
2. time triangulation: involves collecting data on the same phenomenon at different points in time;
3. space triangulation: the collection of data in multiple sites; and
4. person triangulation: involves the collection of data from different levels of persons (individuals; groups such as dyads, triads, families; and collectives, e.g. organizations, communities and institutions).

- *Investigator triangulation* refers to the use of more than one researcher – investigator, interviewer, observer, researcher, data analyst – in an investigation.
- *Method triangulation* refers to the use of multiple methods (interviews, observations) to study a situation or phenomenon.
- *Theory triangulation* refers to the use of multiple theories or hypotheses to examine a situation or a phenomenon.

defined gender from a perspective of family relationships (head of household, husband, wife, decision-making power and control over resources) and communities (participation in organizations and/or groups) (Mehar, 2016). Different levels of measurement, sampling strategies and research methods are applied to these data sources. Different data sources are used. To achieve a complete picture, each data source was chosen to provide data that were potentially different from those gathered from the other two groups of participants. As they are not directly comparable, we use descriptive analysis to provide broad information related to individuals' cognitive behaviour towards climate variability.

In both data sets, the gender of the farmer has been considered in designing surveys. The first author of the book was involved in design of the data sets and has used the information in her academic training (MPhil and PhD theses). Descriptive data used in this chapter are relevant to, and a revised version of, information published in these theses; and the theses are referred to in the results section as the source of descriptive data.

The data sources used are:

CCAFS (2015). This is on the CCAFS baseline data set collected in 2011/12 in three countries – Bangladesh, India and Nepal. Mehar's (2014) MPhil thesis used the data set of 641 farm households in 12 villages of the Vaishali district of Bihar, India. The thesis primarily aims to identify male and female farmers' differences, if any, in perception, knowledge and experience of climate change including the coping strategies adopted by them. The baseline data has used three different gender units: (i) intra-household: gender identity by household member involved in farming activities; (ii) by member involved in decision making; and (iii) some questions asked to both husband and wife.

Yamano (2014–2015). This is a household survey conducted under the Rice Monitoring Survey (RMS) of 6000 households in four eastern Indian states. Mehar's (2016) PhD thesis used the data set to understand farmers' perception, knowledge and experience about submergence and drought in general and across crop stages. It also explores farmers' behaviour in adoption of different varieties including high-yielding varieties and stress tolerant varieties. The gender unit referred to here is the gender identity of the household member who is interviewed and is the prime decision maker in farming decisions.

5.3 Descriptive Statistics

In this section, we use data from the CCAFS (2015) study. Table 5.1 represents respondent farmers' awareness about climate change/variation. More than 90% of the farmers have heard about climate change. Both male and female respondents display a similar pattern of awareness.

To understand their knowledge territory, farmers were requested to mention the different sources from where they heard about climate change (Table 5.2). The major information sources for farmers are reported as: other community members, radio and newspapers. This indicates that the surveyed farmers approach other farmers or institutions to gather updated information about farming.

Table 5.1. Awareness about climate change.

	Male	Female	Total
No	43	9	52
	(8)	(8)	(8)
Yes	479	106	585
	(92)	(92)	(92)
Total	522	115	637

Note: a few farmers who reported that they didn't know about it are not represented here. Figures in parentheses: relative frequency of total farmers who responded.

Table 5.2. Information source – where farmers heard about climate change.

	Male (frequency)	Female (frequency)
Radio	202	21
Television	80	8
Newspapers	168	22
Other community members	284	53
Weather forecasting department	3	0
Local institutions	10	2
NGOs	14	4
Others	60	39

Note: multiple responses were given by farmers, so the total number is more than the sample size.

Female farmers, however, as reported earlier, are mostly illiterate and have limited mobility and therefore tend to gather information from other community members. The formal weather forecasting department is not a vital source of information and efforts should be made to improve the information delivery from this source at local level. It is important to note that these departments may not interact directly with the farming community; rather the farmers get information mainly through mass media such as TV, radio and newspapers.

Farmers were also asked to identify the different causes of climate variation. The results are reported in Table 5.3. The major reasons farmers reported as the cause of climate change are: increasing deforestation, pollution, will of God, and increase in population. Very few farmers reported livestock/agricultural activities as the reason for climate change.

Around 99% of farmers reported that they have felt changes in weather conditions affecting their agricultural activities over the past five to ten years. Almost 75% use weather forecasting information available to them in making farm production decisions (Table 5.4).

Farmers were asked to report changes they observed in the climate/weather over the last 15–20 years. The top three changes that were mentioned by each farmer are shown in Table 5.5. Farmers reported the major observable changes that they relate to climate change or climate variability as higher temperatures, less rainfall and heatwaves.

To understand perceptions about climate change and capture more gender-specific views, farmers were requested to say whether they felt climate change is a problem for them now as well as in the next five to ten years (Table 5.6). Later, the

Table 5.3. Farmers' responses regarding reasons behind climate variation.

	Male (frequency)	Female (frequency)
Increased industrial activity	54	5
Deforestation	410	71
Pollution	282	52
Increase in population	157	32
Livestock/agricultural activities	14	3
Human nature	53	9
Reasons not related to human activity (natural)	28	4
Will of God	244	61

Note: multiple responses were reported.

Table 5.4. Whether farmers use weather forecasting information to assist in making farm production decisions.

	Male (frequency)	Female (frequency)
No	146	24
Yes	375	89
Total	521	113

Note: some farmers did not respond.

Table 5.5. Changes observed in climate/weather over the last 15–20 years.

Changes	Male (frequency)	Female (frequency)
Higher temperature	477	77
Heatwaves	189	38
Less rainfall	453	78
Prolonged drought	66	13
Frequent drought	63	12
Stronger winds/storms	24	4
Hot months getting hotter	160	17

Note: multiple responses were reported.

spouses of the responding farmers were also asked the same question. Their views are categorized into a five-point Likert scale, i.e. strongly agree, agree, neutral, disagree, strongly disagree. More than 90% of the husbands (or male farmers) agreed that climate change is a problem for them now and will be in the next five to ten years. More than 80% of the wives (or female farmers) also agreed. No farmer in our sample *strongly* disagreed with this view, only disagreed.

The respondents were also asked to state the major climate risk affecting their crops (Table 5.7). The results suggest that, for rice crops, the major risks are higher temperature, less rainfall or drought, as well as a reduction in water tables (from others). For wheat crops, the major risks were winter intensity and duration, and shift in timings, and these were also reported as climate risks (Table 5.8). The pattern of responses between male and female for both crops is quite similar.

Table 5.6. Is climate change a problem for farmers now?

	Climate change is a problem for me now		Climate change will be a problem for me in the next 5–10 years	
	Husband (%)	Wife (%)	Husband (%)	Wife (%)
Strongly agree	19.89	28.52	28.91	20.82
Agree	74.45	64.53	51.92	63.34
Neutral	5.47	6.76	19.17	15.84
Disagree	0.18	0.18	0	0
Strongly disagree	0	0	0	0
Total sample	548	547	339	341

Table 5.7. Which climate risks affect crops most?

	Rice		Wheat	
Climate risks	Male	Female	Male	Female
Higher temperature	74.29	57.76	15.24	9.48
Heatwaves	12.76	25.00	3.81	2.59
Less rainfall	83.05	74.14	17.90	13.79
Prolonged drought	19.24	21.55	1.71	1.72
Frequent drought	10.86	12.07	1.52	2.59
Hot months getting hotter	23.05	14.66	4.76	4.31
Late start of rains	12.57	16.38	0.76	1.72
Cold waves	0.00	0.00	20.38	17.24
Frost	0.00	0.00	25.90	37.07
Decrease in duration of winters	0.00	0.00	24.57	37.93
Cold months getting colder	0.00	0.00	11.43	12.93
Shift in timings of seasons	0.00	0.00	29.71	30.17
Others	24.95	25.86	32.38	18.10

Multiple responses: farmers reported three major risks without ranking, so we have used a simple average to calculate the responses.
Note: 'others' include higher intensity of rainfall, frost, shift in timings of seasons, hot months getting hotter, more overall rainfall, more frequent floods, cold waves, average temperature increasing, water table getting lower.

Farmers claimed that the above climate risks will impact their crops by reducing production and crop loss (full or partial, depending on the severity of the risk). Farmers have also reported that with climate risk there is also an increase in pests and diseases as well as the emergence of new types of pests and diseases which harm their crops. Gender-disaggregated results show a similar pattern.

5.4 Climate Stress Perception and Experience with Seasonal Variability

This data set (Yamano, 2014–2015) and analysis (CCAFS, 2015) specifically focus on rice farmers' experience, knowledge and perception of climatic stress in

Table 5.8. Perception about the impact of climate risk on crops.

	Rice		Wheat	
Impact indicators	Male	Female	Male	Female
No effect	4.00	0.00	5.52	0.86
Less production	68.95	73.28	69.71	73.28
Crop loss (full or partial)	42.10	25.00	40.95	21.55
Crop pests (increase and emergence of new pests)	47.05	37.93	45.52	43.10
Crop disease	28.95	23.28	25.71	24.14

Multiple responses: farmers reported three major impacts without ranking, so we have used a simple average to calculate the responses.

the studied season. Farmers' experience, knowledge and perception about climate variation has been evaluated using the following Likert scales (expressing degrees of agreement) or dichotomous questions (yes/no):

1. Do you think climate stress (submergence and/or drought[1]) has become worse/remained the same/decreased?
2. In the past five years, in how many years has submergence/drought been a problem?
3. To what extent do you agree with the statement that submergence/drought is a problem for your plot? (strongly agree/agree/neutral/disagree/strongly disagree).

From Table 5.9 it is clear that for male and female farmers drought is a major problem and salinity less so. Around 50% of male and female farmers have also reported their rice farm being submerged for at least one day over the previous five years. However, the percentage increases to 70% when reporting experience of drought in at least one year. For intensity of drought, as compared to submergence, male and female farmers have a similar pattern of response but not significant differences.

Farmers were also asked about their actual experience during the previous cropping season. On average, 41.85% and 26.71% had experienced drought and submergence, respectively, in at least one of the rice varieties they had sown. Figs 5.1 and 5.2 present the farmers' experiences of submergence and drought by crop stages. It is important to acknowledge that sometimes the stress remained for two or more cropping stages (Yamano, 2014–2015). Fig. 5.1 shows that farmers, on average, experience submergence mostly during tillering, and panicle-initiation to heading–flowering stages of the rice crop, whereas they reported experiencing drought mostly during sowing, tillering and panicle-initiation stages (Fig. 5.2). Male and female farmers had similar responses, but the results are not presented here.

5.5 Summary

The research described in this chapter provides insights into how the questions related to climate change and gender are asked and assessed. The authors' aim

Table 5.9. Farmers' perception and experience about climate stresses. (From Yamano, 2014–2015)

	Submergence		Drought		Salinity	
	Male	Female	Male	Female	Male	Female
Perception about climate stress						
Strongly disagree	31.87	34.73	14.48	14.50	67.00	70.99
Disagree	19.22	19.08	14.95	12.21	20.44	20.23
Neither	4.71	3.82	3.92	4.96	3.95	1.91
Agree	34.11	32.82	43.08	43.51	7.06	3.05
Strongly agree	10.10	9.54	23.59	24.81	1.54	3.82
Experience (in the last 5 years)						
None	51.45	53.05	31.29	29.77	92.11	92.75
1 year	15.99	13.36	11.82	15.65	2.21	0.76
2 years	19.51	19.85	23.98	15.65	3.61	3.82
3 years	7.42	8.02	20.10	20.61	1.10	0.76
4 years	3.90	2.29	7.99	8.02	0.25	0.38
5 years	1.74	3.44	4.81	10.31	0.72	1.53
Severity of problem in the past 10 years						
Getting worse	15.47	11.45	30.48	26.72	2.26	1.53
Staying about the same	23.94	24.05	32.75	33.21	8.93	6.87
Getting less problematic	24.88	16.79	12.47	11.45	16.65	5.73

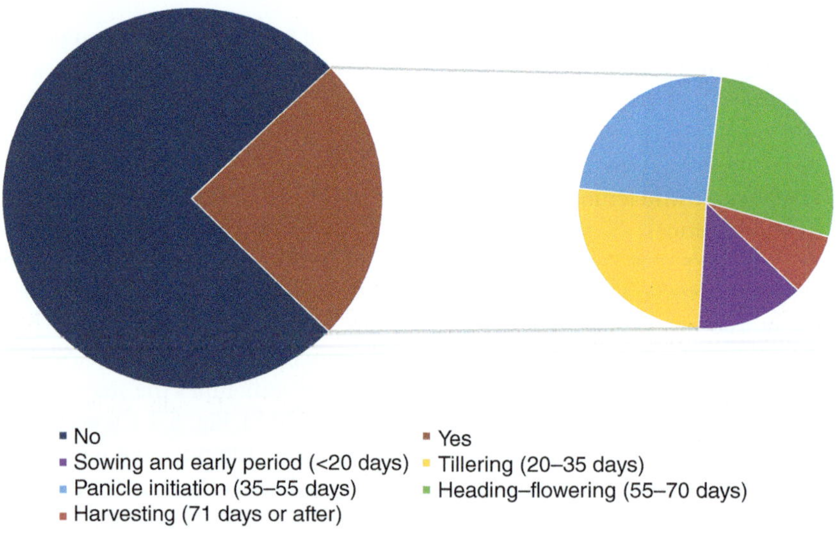

Fig. 5.1. Percentage of farmers experiencing submergence and by crop stages. (From: the author)

is not to focus on the findings but to illustrate the fundamentals of perception, experience and knowledge. The chapter also echoes the recent call by social scientists to investigate perception and experience of climate stress by crop stages and recommends considering multi-stress situations. Lastly, the chapter has suggested the value of triangulation methods, which offer utilization of secondary data or information to understand the research context and problem. In the case of

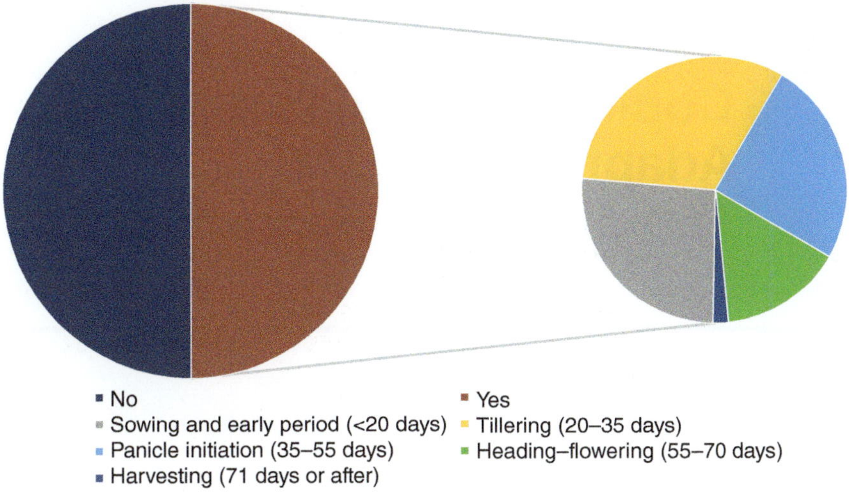

Fig. 5.2. Percentage of farmers experiencing drought and by crop stages. (From: the author)

inconsistent or contradictory evidence from triangulation methods, the researchers are encouraged to examine the psychological barriers to the acknowledgment of climate change and action for solutions.

Note

[1] Note: the term 'drought', as used here, indicates the number of days when there was no rainfall.

References

Bryman, A. (2009) Mixed methods in organizational research. In: Buchanan, D.A. and Bryman, A. (eds) *Handbook of Organizational Research Methods*. Sage, London, pp. 516–531.

CCAFS (2015) Household Baseline Survey 2010–2012. Available at: https://doi.org/10.7910/DVN/IUJQZV (accessed 8 April 2022).

Denzin, N.K. (1989) *The Research Act*, 3rd edn. McGraw-Hill, New York.

Denzin, N.K. (2006) *Sociological Methods: A Sourcebook,* 5th edn. Aldine Transaction, Piscataway, New Jersey.

Mehar, M. (2014) Gender specific impacts of climate variation in agriculture. Unpublished MPhil thesis, Department of Economics, School of Social Science, Indira Gandhi National Open University, India.

Mehar, M. (2016) Role of gender differentiated risk and time preferences in rice varietal selection in eastern India. PhD thesis, Department of Economics, School of Social Science, Indira Gandhi National Open University, India. Available at: https://shodhganga.inflibnet.ac.in/handle/10603/212284 (accessed 8 April 2022).

Otzelberger, A. (2011) *Gender-Responsive Strategies on Climate Change: Recent Progress and Ways Forward for Donors*. BRIDGE, Institute of Development Studies, University of Sussex, Brighton, UK.

Yamano, T. (2014–2015) *Rice Monitoring Survey: South Asia*. Available at http://ricestat.irri.org/fhsd/php/panel.php (accessed 8 April 2022).

6 Coping Strategies and Adaptation Strategy: Framings and Perspectives

Abstract

This chapter elucidates the concepts and their dimensions that are needed to define and understand how people respond to and adapt to climate impacts. The two distinct, but inter-related, concepts for responses are adaptation and mitigation. The undeniable fact is that climate hazards have a negative effect on agricultural production as well as on food access nutrition, especially for households that rely upon agriculture for their livelihood.

6.1 Introduction

> 'While the causes of climate change are known, i.e., both nature- and man-induced activities, it can be concluded that efforts of mitigation should become a norm.'
>
> (Zwane, 2019)

From the previous chapter it is clear that climate change is an issue that is affecting human societies and it is often difficult to disentangle the role of and impact on males and females. Gender studies remain focused on women as the major victims of climate change impacts because of their role and dependence on natural resources. Ironically, literature, though relatively few, also shows them to be the first observers as well as the effective agents of change, as they often use their indigenous knowledge to cope with and adapt to climate change (Israel and Sachs, 2013). Women often have a strong body of knowledge and expertise that can be used in climate change mitigation, disaster reduction and adaptation strategies. Like other aspects of climate change, how people respond is not gender-neutral, and they respond in variable ways and with varying actions at local level. This variability is due to the different needs and experiences of women and men, and girls and boys, and their capacity to address climate change.

The first few searches on 'how people respond to and adapt to climate impacts' uncovered words such as: coping, adaptation, mitigation, climate smart intervention,

DOI: 10.1079/9781789249910.0006

climate resilience, risk assessment, etc. Clarification of these terms, along with intervention and available strategies to adapt to climate change, is the focus of this chapter.

6.2 Adaptation and Mitigation

Mehar (2014) has divided farmers' choices in dealing with climate risks into two categories: generic (education, income, health, additional job, etc.) and specific (crop rotation, early planting, irrigation, adoption of hybrid seeds, etc.) (Table 6.1).

These responses are scientifically defined differently based on their action type, timing and scale. The two-pronged approach used as a response to climate change is adaptation and mitigation. Another expression that is used is 'coping', which is often confused, and used interchangeably, with 'adaptation'. Although they are different terms, the difference can be viewed with 'spatial scale' and 'timescale'. Coping is typically used to refer to *ex-post* actions, while adaptation is normally associated with *ex-ante* actions (Cardona *et al.*, 2012, p. 72). Other distinguishing features of coping and adaptation are that coping is orientated towards survival and reactive strategies that are not continuous, whereas adaptation practices are planned for longer-term improvement of production and livelihood.

Coping strategies are usually short-term activities to deal with climate change and might even have negative impacts in the medium or long term (Cardona *et al.*, 2012). The respective actions here are rooted in favour of one gender. Amartya Sen's entitlement theory (Sen, 1981, 1999) draws attention to action decisions rooted in gender inequalities in food acquisition within families in these climate change-affected regions. '*Consumption smoothing* or the adjustment of food consumption within the household, as outlined by Sen, is facilitated by women eating less, a strategy adopted in many households to address food insecurity' (Alston and Akhter, 2016). Income smoothing for food security is rooted as the responsibility of men; often the actions of borrowing and migration are seen as masculine traits.

Coping strategies depend on coping capacity, i.e. the ability to react to and reduce the adverse effects of experienced hazards. They are often considered part of

Table 6.1. Categorization of farmers' choices. (From Mehar, 2014)

IPCC indicators	Specific category examples
Generic indicators	**Borrowing** (from friends, relatives, money lenders, commercial banks, Kisan credit card) Eating less (reduced meal size/quality, buying food from market) Additional job (non-agricultural in village/out-migration)
Specific indicators	**Crop rotation** Use of hybrid seeds Other agricultural practices (replanting, reduced use of chemical fertilizer, use of manure, use of pesticides/herbicides, diversification of crops, substitute crops, intercropping, changing planting date, water harvesting, use of organic manure, land levelling)

adaptive capacity, which means the ability to anticipate and transform structure, functioning or organization to better survive hazards (Saldaña-Zorrilla, 2007).

The other category of 'specific indicators' includes *ex-ante* approaches and these are selected based on farmers' uncertainty about the weather for the cropping season and cropping stages (from sowing to harvesting). The approaches can be divided into two broad categories: adaptation and mitigation. Adaptation and mitigation responses to climate change share the same final target of sustainable development yet are highly differentiated from social, spatial and temporal scales (Table 6.2).

Refer to the glossary section for other related concepts such as adaptation opportunities, adaptation constraints, adaptation limits (soft and hard), etc.

Adaptation or adaptation-related responses can include a range of responses to climate change. There are three broad categories of adaptation based on the intent of implementation (autonomous or planned), timing of implementation (anticipatory or reactive) and degree of change (transformational or incremental):

Autonomous or planned. Autonomous or spontaneous adaptation is an immediate reaction to ecological changes such as a change in planting date, using location-specific tolerant varieties or diversifying income sources. Planned adaptation, as the name suggests, includes conscious intervention or preparation by a farmer as well as government intervention, such as investment in infrastructure, subsidies and tax regime, innovation and research.

Anticipatory or proactive and reactive adaptation. Reactive adaptation occurs after the effects of climate change are evident. In contrast, anticipatory or proactive adaptation occurs before the effects of climate change are evident.

Incremental, system or transformative adaptation.

> Incremental adaptation refers to actions where the central aim is to maintain the essence and integrity of the existing technological, institutional, governance, and value systems, such as through adjustments to cropping systems through new varieties, changing planting times, or using more efficient irrigation. In contrast, transformational adaptation seeks to change the fundamental attributes of systems in response to actual or expected climate and its effects, often at a scale and ambition greater than incremental activities. It includes changes in activities, such as changing livelihoods from cropping to livestock or by migrating to take up a livelihood elsewhere, and also changes in our perceptions and paradigms about the nature of climate change, adaptation, and their relationship to other natural and human systems.
>
> (Noble *et al*., 2014, p. 839)

Fig. 6.1 presents the adaptation strategies of reducing the impact of climate change on social-ecological systems, with examples from agriculture, along a gradient of increasing magnitude of responses.

The adoption decision of adaptation strategies is influenced by adaptive capacity and risk assessment. Both of these elements are scale- and place-specific and thus vary across nations, communities and among social groups and individuals.

6.3 Risk Assessment

Effective implementation of mitigation and adaptation strategies requires proper accounting for uncertainty and risk. The IPCC, in its fifth Assessment Report

Table 6.2. Adaptation versus mitigation. (Adapted from various IPCC documents)

	Adaptation	Mitigation
The aspect of climate change	Adaptation deals with the effects and impacts of climate change.	Mitigation deals with causes of climate change.
Definition	IPCC definition: 'an adjustment in natural or human systems in response to actual or expected climatic stimuli or their effects, which moderate harm or exploit beneficial opportunities' (IPCC, 2001). Other definition: cultural or behavioural practices that allow societies to minimize effects related to climate risk.	IPCC definition: 'an anthropogenic intervention to reduce the emissions sources or enhance the sinks of greenhouse gases' (IPCC, 2011). Other definition: measures taken to reduce and stabilize the adverse impacts of hazard, exposure and vulnerability.
Spatial scale	Primarily a local issue and provides a benefit (immediate) at a local level. Immediate benefit(s) can be expected though adaptation usually is not able to eliminate all negative impacts of climate change.	Primarily an international issue, as mitigation provides global benefits. Lagged but long-term benefit with a limit or eliminating changes in the climate system can be expected.
Timescale	It has a short-term effect on the reduction of vulnerability.	It has a long-term effect because of the inertia of the climatic system.
Equity	'Victim' not always responsible for causing climate change.	Especially motivated with countries less vulnerable to climate change.
Evaluation	Monitoring: no defined 'goal' or indicators to measure adaptation. Cost and benefit: much more difficult to calculate (especially benefits).	Monitoring: relatively straightforward to measure emissions and compare to targets. Cost and benefit: reasonably straightforward to determine, compare and aggregate.
Types of parameters	Types: anticipatory and reactive, private and public, and autonomous and planned.	Components: CO_2, N_2O, CH_4 Parameters: energy efficiency, emissions, carbon sequestration.
Examples	Related to crops: introduce drought-tolerant varieties, use of precision agricultural techniques, water management practice (e.g. rain harvesting, drip irrigation). Related to livestock: improved feed management.	Related to crops: use of precision agriculture techniques, plant management (e.g. cover crops). Related to livestock: improved feed and improved breeds of livestock reduce emissions from enteric fermentation (CH_4).
Synergies	Improved-input carbon practices (improved crop varieties, crop rotation, use of cover crops, perennial cropping systems, agricultural biotechnology); improved N_2O efficiency. Integrated or mixed production systems increase land productivity and efficiency in use of water and other resources and protect against soil erosion as well as serve carbon sequestration objectives.	

(AR5), introduced a risk-based framework to assess adaptation (IPCC, 2014). This framework defines risk as a combination of three elements: hazard, vulnerability and exposure. Since then, there is increasing interest in understanding and assessing risk to understand decision-making adaptation. The key defining feature of risk in the climate change context is uncertainty about weather events in the future and adaptive capacity. The risk may be low or high if the adaptive capacity of an individual is high or low, respectively.

Building on the risk concept introduced in the IPCC's AR5, and presented in Fig. 6.2, the gender dimension of climate analysis can be examined. Natural hazards, both slow- and rapid-onset, are gender-neutral; their intensity remains the same for all ages of men and women. However, the impact of any disaster – its type, intensity and scale – depends on who is exposed and the level of vulnerability,

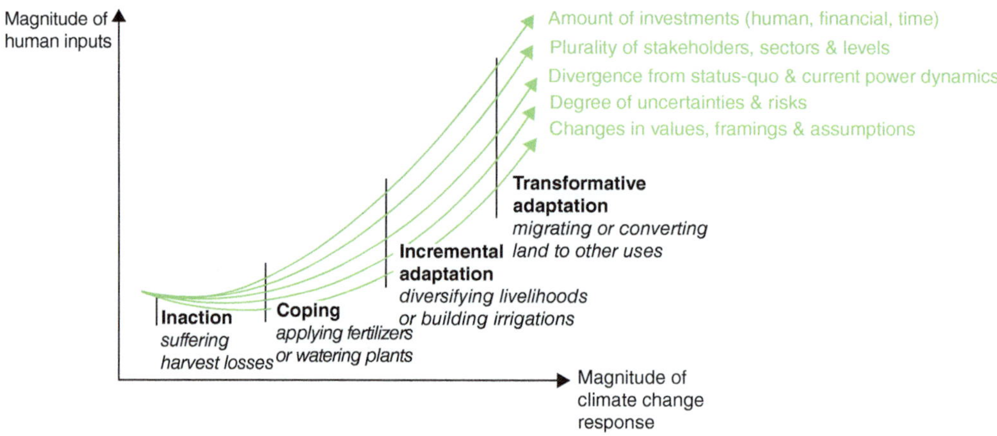

Fig. 6.1 Adaptation strategies. (Adapted, with permission, from Fedele *et al.*, 2019)

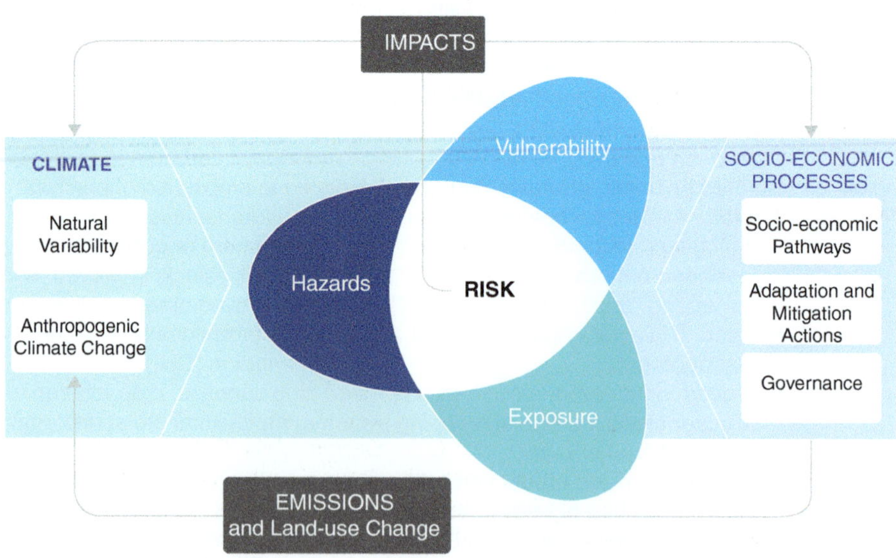

Fig. 6.2 The IPCC AR5 conceptual framework with risk at the centre (IPCC, 2014).

adaptive capacity or preparedness to face or recover from the event, and this may be different for men and women and may vary across societies, based on the social norms and levels of agency embedded in the society. Gender-differential access to and control over resources influence gender-specific vulnerability as well as the way men and women are exposed to climate change impacts. The differentiated impacts, power and adaptive capacity define the potential for risk taking in the particular context.

However, risk-tolerance levels can be improved by creating a supportive environment for farmers. Regarding the existing environment and potential strategy to enhance farmers' risk capacity, the existing factors affecting risk tolerance have three prime aspects: (i) context-ecological and social conditions; (ii) individual capability and capacity: individual socio-economic characteristics; and (iii) policy and networking: stakeholders' nature and extent of involvement towards climate-friendly technologies and innovations. This underscores the importance of multiple climate change risks and their interactions. Simpson *et al.* (2021) has provided a framework to clarify the types of interactions that generate risk and where they originate.

6.4 Summary

This chapter has distinguished concepts of adaptation and mitigation: the two critical approaches used in response to climate change. It also discussed how and why risk assessment is important for adaptation strategies.

References

Alston, M. and Akhter, B. (2016) Gender and food security in Bangladesh: the impact of climate change. *Gender, Place and Culture* 23(10), 1450–1464.

Cardona, O.-D., van Aalst, M.K., Birkmann, J., Fordham, M., McGregor, G. *et al.* (2012) Determinants of risk: exposure and vulnerability. In: Field, C.B., Barros, V., Stocker, T.F. and Qin, D. *et al.* (eds) *Managing the Risks of Extreme Events and Disasters to Advance Climate Change Adaptation.* A Special Report of Working Groups I and II of the Intergovernmental Panel on Climate Change (IPCC). Cambridge University Press, Cambridge, UK, and New York, pp. 65–108.

Fedele, G., Donatti, C.I., Harvey, C.A., Hannah, L. and Hole, D.G. (2019) Transformative adaptation to climate change for sustainable social-ecological systems. *Environmental Science and Policy* 101, 116–125.

IPCC (2001) *Climate Change 2001: Synthesis Report.* A Contribution of Working Groups I, II, III to the Third Assessment Report of the Intergovernmental Panel on Climate Change (R.T. Watson and the Core Team (eds)). Cambridge University Press, Cambridge and New York.

IPCC (2014) *Climate Change 2014: Impacts, Adaptation, and Vulnerability. Part A: Global and Sectoral Aspects* (edited by Field, C.B., Barros, V.R., Dokken, D.J., Mach, K.J., Mastrandrea, M.D. *et al.*). Cambridge University Press, Cambridge, UK and New York.

Israel, A.L. and Sachs, C. (2013) A climate for feminist intervention: feminist science studies and climate change. In: Alston, M. and Whittenbury, K. (eds) *Research, Action and Policy: Addressing the Gendered Impacts of Climate Change*, pp. 33–52.

Mehar, M. (2014) Gender specific impacts of climate variation in agriculture. Unpublished MPhil thesis, Department of Economics, School of Social Science, Indira Gandhi National Open University, India.

Noble, I.R., Huq, S., Anokhin, Y.A., Carmin, J., Goudou, D. *et al.* (2014) Adaptation needs and options. In: Field, C.B., Barros, V.R., Dokken, D.J., Mach, K.J., Mastrandrea, M.D. *et al.* (eds) *Climate Change 2014: Impacts, Adaptation, and Vulnerability. Part A: Global and Sectoral Aspects*. Cambridge University Press, Cambridge, UK and New York, pp. 833–868.

Saldaña-Zorrilla, S.R. (2007) *Socioeconomic Vulnerability to Natural Disasters in Mexico: Rural Poor, Trade and Public Response*. CEPAL Report 92, UN-ECLAC, Disaster Evaluation Unit, Mexico, ISBN: 978-92-1-121661-5.

Sen, A.K. (1981) *Poverty and Famines: An Essay on Entitlement and Deprivation*. Clarendon Press, Oxford, UK.

Sen, A.K. (1999) *Development as Freedom*. Knopf, New York.

Simpson, N.P., Mach, K.J., Constable, A., Hess, J., Hogarth, R. *et al.* (2021) A framework for complex climate change risk assessment. *One Earth* 4(4), 489–501.

Zwane, E.M. (2019) Capacity development for scaling up climate-smart agriculture innovations. In: Hussain, S. (ed.) *Climate Change and Agriculture*. IntechOpen, London.

7 Gender Inclusiveness in Climate-smart Agriculture

Abstract

This chapter describes the conceptual understanding of 'climate-smart agriculture'. It provides evidence from South Asia and Africa to unpack the gender-inclusiveness in CSA approaches.

7.1 Climate-smart Agriculture

Climate-smart agriculture (CSA) aims towards a triple win: in addition to adaptation and mitigation it includes food security goals. Introduced in 2010, the CSA concept is defined as agriculture that sustainably increases productivity and resilience (adaptation) and reduces or removes greenhouse gases (mitigation) while enhancing the achievement of national food security and development goals (FAO, 2010). This concept is supported and promoted by several international multilateral agencies such as the World Bank, the International Fund for Agricultural Development (IFAD) and the Consultative Group on International Agricultural Research (CGIAR) (FAO, 2010, 2013). Several definitions of the CSA concept have emerged from the literature (Collins-Sowah, 2018), albeit remaining around the three main pillars (Fig. 7.1) of CSA:

- adapting and building resilience to climate change;
- reducing and/or removing greenhouse gas emissions, where possible; and
- sustainably increasing agricultural productivity and incomes.

The CSA conceptual framework documented in Collins-Sowah (2018) depicts all the relevant linkages. The large, dark-green circles represent the three main pillars and the red intersecting lines provide room for potential synergies and trade-offs within and across these pillars. The green rectangles are adjacent to climate-smart policies, institutions and financing mechanisms (with relative parameters

 Climate Change, Adaptation and Gender: Policy, Practice and Methodological Underpinnings (M. Mehar and N. Prasad)
DOI: 10.1079/9781789249910.0007

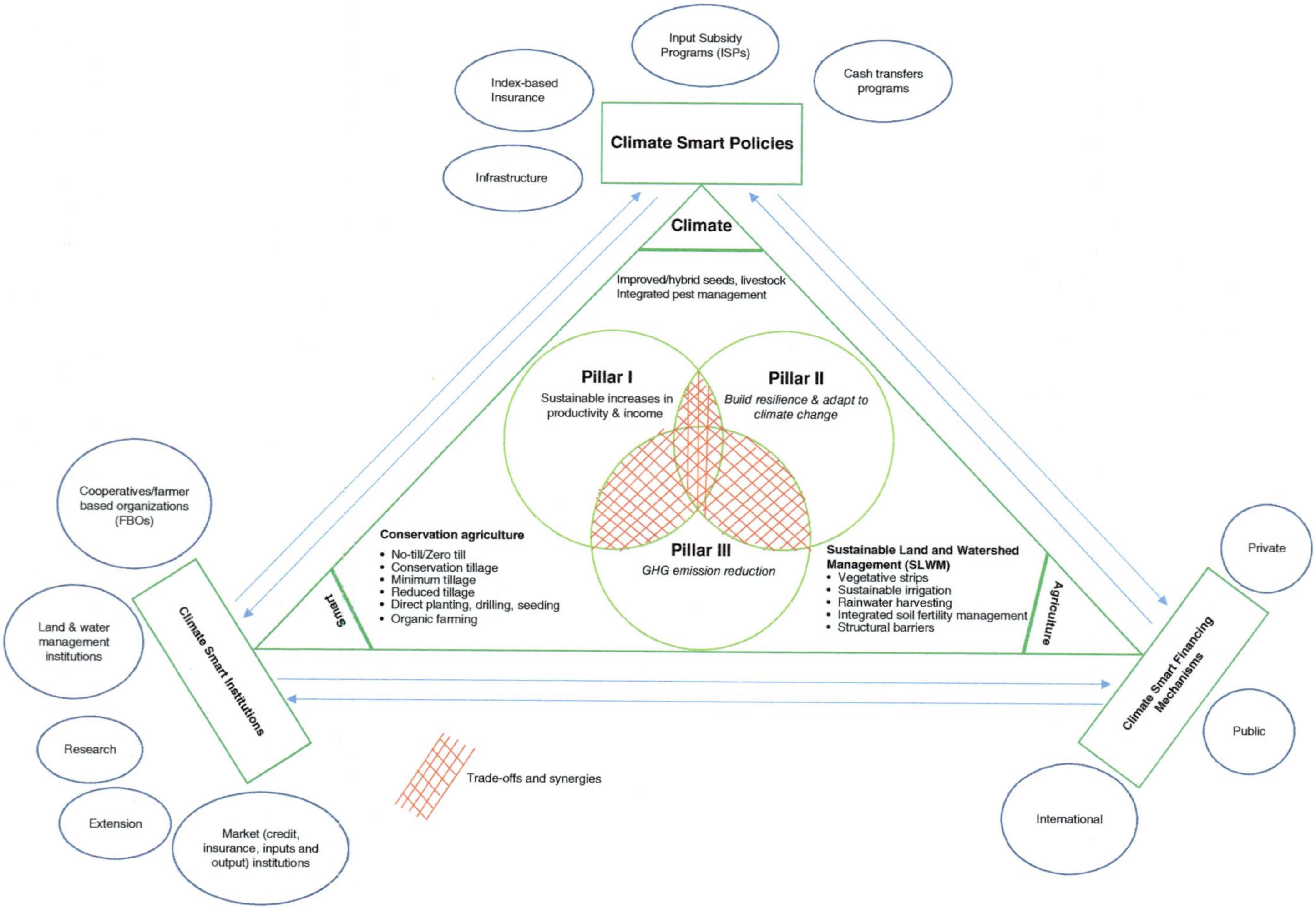

Fig. 7.1. Climate-smart agriculture (CSA) conceptual framework. (Adapted from Collins-Sowah, 2018, with permission)

in the blue circle) which serve as enabling and supporting factors that facilitate the planning and implementation of CSA programmes and also the adoption of CSA technologies and management practices by farmers. The size of the intersecting areas among the various pillars varies depending on the technology set chosen, location, time and socio-economic context.

Few studies have shown that selection and adoption of CSA practices can help in achieving larger benefits at least cost and have proved advantageous business models to farming communities among many localities (Schroth *et al.*, 2015; Williams *et al.*, 2020). Ng'ang'a (2021) found that, except for minimum tillage, other CSA practices have positive net present value (NPV) and a larger internal rate of return (IRR) than the discount rate for their lifespan.

The other key attributes of CSA are:

- It is an 'approach for transforming and reorientating agricultural development under the new realities of climate change' (Lipper *et al.*, 2014).
- 'CSA as a concept goes beyond agricultural practices and technologies to include enabling policies and institutions as well as identification of financing mechanisms' (FAO, 2013).
- It is not a one-size-fits-all and should be conceived as context-specific, an evolving process that needs to be prioritized and tailored based on the type, frequency, timing and intensity of climate hazards.
- There are potential trade-offs and synergies within the three main goals of CSA, in the integration of agricultural sectors and other associated sectors (such as energy, water and market). (More discussion with examples (with gender aspects) around these are provided in the next chapter.)

Achieving the triple-win outcomes of CSA is very difficult, especially in regions where they are needed most, regions that are resource-poor and subject to the vagaries of nature. So planners have to make trade-offs and strategies identifying the synergies and weighing the costs and benefits of CSA practices. Below are a few examples of synergies when the three outcomes occur simultaneously with a positive effect on each other:

System of rice intensification (SRI). It uses less water (adaptation and resilience), reduces methane emission (mitigation), increases yield and hence income and food security.

Agroforestry, especially growing nitrogen-fixing trees. It can lead to mitigation (due to soil and biomass carbon sequestration, buffers heat stress and erosion), adaptation and resilience to climate extremes (by enriching soil fertility and soil-water retention) and increased sustainability and income.

Integrated system. An integrated rice and livestock system is a mitigation strategy (reduction of methane emissions) with co-benefits of adaptation and resilience (animal manure utilization for soil fertility) and is sustainable (diversification of income). Similar benefits can be viewed from another integrated system: agroforestry and silvopastoral systems, crop–aquaculture (rice–fish farming), etc.

Other than the sustainable increase in production and income, some CSA activities also assure other benefits. For example, the use of biogas for cooking stoves over traditional wood-based ones has benefits for health and for reducing time poverty. Some stress-tolerant varieties, in addition to being resilient to climate variability (e.g. drought, salinity, submergence), also have other traits. For example, an orange-fleshed sweet potato variety disseminated in 14 sub-Saharan African countries has traits of drought-tolerance and high pro-vitamin A content (Low *et al.*, 2017).

Like the above synergies and co-benefits, there are also trade-offs and negative effects among the three pillars associated with some strategies. For example, sustainable agricultural intensification is used to avoid the constraint of limited land and has little or no impact on mitigation due to lack of carbon-emitting technology used in intensification.[1] Evidence suggests that sustainable intensification could lead to an increase in absolute greenhouse gas (GHG) emissions. For example, with multi-crops, soil-organic carbon declines due to continuous cultivation of crops. Another example of adaptation strategy common in climate extremes is livelihood diversification. Suckall *et al.* (2014) suggested diversification options such as tourism activities will increase carbon, which can be seen in emissions from flights and hotels. The other associated concern of diversification is that males look for outside work opportunities, which increases women's time burden in doing activities related to farming that were previously done by men.

7.2 Smart in Climate-smart

CSA has multiple entry points, and the operationalization or selection of CSA options is based on technical appropriateness, climate sensitivity, scalability and farmers' acceptance. *Inter alia* benefits of CSA technologies, practices and services can be viewed in terms of energy-, water-, carbon-, nutrient-, knowledge-, weather-, and gender and social inclusion-smart. These are discussed below with examples.

Energy-smart. Interventions to support environmentally sustainable energy sources. These interventions aim to increase energy-use efficiency, reduce or minimize use of fossil fuel-based technologies (e.g. zero tillage, drip irrigation, combine harvester, energy-efficient cooking stoves), promote the use of biofuels (e.g. biogas) and renewable energy sources (e.g. solar irrigation).

Water-smart. Interventions aimed at enhancing the efficient use of water. This could be done in several ways: increased water efficiency by reusing, recycling and reducing or optimizing use of available water (e.g. community management of water, aquifer recharge, drainage management, plastic ponds for rainwater or waste water management); addressing water scarcity by using improved land-preparation methods (e.g. laser land levelling, broad-bed furrow, raising crops on beds); irrigation methods (e.g. drip irrigation, rainwater harvesting, micro-irrigation); increased use of resilient crops and varieties to address uncertainty of water or monsoons (e.g. cover crop or stress-tolerant crop varieties); and minimization of water loss from the surface of fields (e.g. vertical drainage system). Patle *et al.* (2019) have discussed all the water-smart technologies for national and international applicability

Carbon-smart. Interventions emphasizing reducing emission of GHGs and increasing carbon sequestration. Interventions to reduce GHG emissions are livestock and manure management and, more specifically, use of concentrate feeding of livestock to reduce methane production. Interventions to increase carbon content in the soil could be done by manure and crop residue management (e.g.

green manuring, mulching, fodder management, use of biochar); by adapting integrated practices or diversifying land-use systems (such as agroforestry, planting perennials, fruit orchards); by supporting reduction in use of fossil fuel-based technologies (e.g. zero/minimum tillage); and by improved practices (e.g. integrated pest management).

Nutrient-smart. Interventions with double objectives: interventions aiming to improve the soil nutrients as per the crop and land requirement and interventions improving the nutrition in farm products to improve nutrition intake of people. These interventions include intercropping with legumes, green manuring, crop rotation, use of organic or bio-fertilizer and bio-pesticide, leaf-colour charts, hand-held crop sensors, soil conservation and integrated hedgerows, improved cattle sheds and farmyard manure, nitrogen fixation by intercropping, site-specific nutrient management.

Knowledge-smart. Interventions aimed at risk management (e.g. crop and livestock insurance, soft loan facilities, diversification of crops, stress-tolerant varieties (which are tolerant to drought, salinity, submergence, flood, heat/cold stresses, changes in planting time); capacity building (e.g. climate field school, promoting farmer-to-farmer learning, building community seed and fodder banks); ICT-smart information, which connects science to local needs with ICT-based agro-advisory services (e.g. market information, weather information).

Weather-smart. Interventions to help farmers minimize loss due to unfavourable weather conditions. This includes providing weather information (e.g. forecasting rain, hailstorms, etc.) most often using ICT-smart channels; creating infrastructure to withstand climate shocks (e.g. tunnel for hailstorms and rains, check dams, hailstone net); transferring weather-related risk to third parties (e.g. weather index-based crop or livestock insurance); farmers' capacity building (creating social safety net); and climate diary (where they can monitor changes in weather and make appropriate decisions; climate analogues).

Gender equality and social inclusion (GESI)-smart. These interventions are designed to meet context- and user-specific needs. For example, promoting technologies with the aim to reduce women's and girls' workload and drudgery in fetching water from distant places. These include rainwater harvesting, community pond, plastic pond and multi-use water systems. Other interventions are user-friendly technologies or tools (such as rake, hand-held corn sheller, mechanical rice transplanting, cono-weeder); improved cattle sheds to reduce women's workload in cleaning and managing sheds; and promoting home vegetable gardens. Box 7.1 specifies some of the GESI-smart technologies designed with a women-centric approach.

7.3 Gender and CSA

The design, demonstration methods and know-how for the use of CSA practices set an environment which may be positive and may set limits for others, with variation across different sections of society as well as by age and gender. Mehar (2020) explores gender dimensions in CSA practices with a spill-over effect on both gender and present caution and concern for including gender dimensions in CSA uptake. We strongly recommend reading Mehar (2020) for detail. Here we

Box 7.1. Women-centric CSA.

Mangrove restoration project in Guyana. Women made up 80% of the participants in mangrove cultivation, restoration and management. These women organized and formed many volunteer groups (village mangrove action committees) to continue the upkeep and conservation of the mangrove forests. Between 2010 and 2013 the project planted over 460,000 black mangrove seedlings, restoring 10 km or 48 ha of coastal mangrove forest.
Geothermal energy and its residuals in El Salvador. Women from 15 communities are taking advantage of these opportunities, using water condensation to grow and sell roses; heat residuals to dehydrate fruits; and reservoirs – a result of well drilling for the plants – for fishing.
Solar market gardens in Benin. The solar-powered pump drip-irrigation system has reduced water waste by 50% and has relieved women and girls from the time-consuming burden of hand-watering crops.

describe the narratives of the chapter with more examples in the dimension of trade-offs and synergies across CSA pillars.

Women and men have different roles across the cropping stages and have different adaptive capacities and resource constraints due to social norms. Their perception of CSA and adoption may not be the same as for men in their family or community. Additionally, CSA interventions, by design, vary in terms of resource and labour requirements (Mehar, 2020), though recently there have been efforts made to design women-friendly technologies. According to Mehar (2016), women-friendly technologies can broadly be classified into two areas:

- those that are purposely designed and developed to ease the hardship involved in performing agricultural operations that are usually done by women (e.g. garden seeding, fertilizer spreading, cono-weeding, etc.); and
- those that take women away from performing operations consisting of drudgery (e.g. mechanical rice transplanting, mechanical paddy threshing, maize shelling, etc.).

A common woman-friendly technology is the use of biogas cooking stoves or solar cookers. Since, culturally, women are viewed as the natural custodians of food security and cooking in the household, this technology is more likely to be accepted by researchers and households. It is also a good example of synergy (with positive feedback) between all three pillars of CSA.

Technology also has a significant effect in reducing problems such as deforestation, women's and children's time poverty, and respiratory diseases resulting from indoor air pollution (Aguilar, 2010). There are examples of how technology has (Ragasa, 2012; Bernier *et al.*, 2013) and has not (Tsephel *et al.*, 2010) worked for women and men where their needs and preferences have or have not been included in the technology choice and project design. Similar implications for varieties are found in studies by Paris *et al.* (2008). She mentioned that 'ignoring women's indigenous knowledge and preferences for rice varieties may lead to slow adoption of new varieties'. Influenced by this perception, recently efforts have been made to understand and include gender-differentiated perceptions and requirements for stress-tolerant varieties (see, for examples, Mehar *et al.* (2017) for rice; Gebre *et al.* (2019) for maize; and Gilligan *et al.* (2013) for potato).

Quite often, the technologies are designed according to men's needs and physical strength to use them – the 'adoption of advanced management technologies like leaf colour charts [and] canopy sensors are not even examined with a gender-dimension' (Mehar, 2020). Other examples, such as zero tillage (ZT) and laser land levelling (LLL), are heavy and use expensive technology, and, by default, require financial decisions to be taken. Women are often viewed as victims in discourse and have limited access to and control over resources. In addition, bound by social norms, they often face mobility constraints which limit them in approaching providers of these technologies. However, females found ZT (Nyasimi *et al.* (2016) in Tanzania; Aune *et al.* (2006) in Ethiopia) and LLL (Magnan *et al.* (2015) in India) more convincing. The reasons are related to social norms which suggest that ploughing with oxen is difficult and culturally unacceptable for women, so they find using these technologies convenient with the help of a service provider, without breaking social norms.

Like women, men are also perceived to not adopt a few CSA technologies they believe are related to women's work and they will be subject to gossip if they do. For example, in Malawi, intercropping of legumes and maize, which improves soil nitrogen, is viewed to be ineffective when aimed at male agricultural decision makers who view legumes as 'women's crops' (Pircher *et al.*, 2013; Mutenje *et al.*, 2016).

Many technologies and practices in CSA require up-front investment and also medium to large landholdings. Women are often disadvantaged in these due to their poor access to assets including tenure rights, networks to get credit and access to medium or large landholdings.

Like limitation to land rights, women's access and control over resources, compared to men's, discourages them from investing. This is a strong reason, *ceteris paribus*, why women are less likely to adopt CSA practices in a generic situation. Several studies support this argument. For example, studies by Kinkingninhoun-Mêdagbé *et al.* (2008) and Yokouchi and Saito (2016) found that 'women's adoption of improved seeds or irrigation schemes does not always result in the longer-term benefit of an increase in crop yield or income, due to limited access to land, equipment, and markets'.

There are also spill-over (negative) effects on another gender even if they do not adopt the technology (Table 7.1). For example, use of plastic drum seeders have resulted in the loss of hired poorer and landless households (Paris and Chi, 2005). The use of ZT and direct seeded rice (DSR) may lead to increased weeding. As mentioned, these technologies are more likely to be used by men, saving women time and sparing them from drudgery in activities related to transplanting, but cleaning the land has an indirect effect of increased weeding activities. Examples from Nepal and Ethiopia support this argument (Halbrendt *et al.*, 2014; Vandercasteelen *et al.*, 2018).

Another example relates to shifting control over profit from improved milk-production schemes (Fisher *et al.*, 2000). Sain *et al.* (2017) document that the 'use of CSA practices can improve employment of women at the farm and aggregate levels and also that practices can improve the gender labour balance'. The authors further recommend the promotion of gender-sensitive CSA initiatives

Table 7.1. Gender considerations of various CSA interventions in South Asia and Africa. (From Mehar, 2020)

Categories of CSA	Key climate-smart interventions[i]	Examples of gender-differentiated revelation to CSA
Soil management	Zero-tillage (ES, WS, CS)	Males prefer more than women in Tanzania (Nyasimi *et al.*, 2016)
		Appealing to women as ZT means less use of oxen in Ethiopia (Aune *et al.*, 2006)
	Laser land levelling: to ensure uniform water distribution (WS)	Significant gender-differentiated effect of network on households' demand on LLL in India (Magnan *et al.*, 2015)
Crop management	Crop rotation, inter-cropping, on-farm tree planting (NS, CS)	In India, crop rotation was selected by 60% of the male decision makers in the household (Mehar *et al.*, 2016)
		In African countries (Cameroon, Tanzania, Ethiopia), women were found to be less engaged in adaptation practices based on tree plantations (Deressa *et al.*, 2009; Below *et al.*, 2012; Molua, 2012)
		Intercropping preferred by more women than men (Nyasimi *et al.*, 2016)
		In Ghana, gender-responsive participatory approaches reveal that men focus more on changing staple crop varieties, introducing new tree crops, intercropping and rotation with legumes (Naab and Koranteng, 2012)
	Improved high-yielding/stress-tolerant variety of seeds (WS, KS)	Gender-differentiated preferences and adoption for rice (Mehar *et al.*, (2017), maize (Gebre *et al.*, 2019), orange-fleshed sweet potatoes (Gilligan *et al.*, 2013), potato (Gilligan *et al.*, 2013) and drought-tolerant leguminous trees (Bernier *et al.*, 2013)
	Direct seeded rice (WS)	DSR weeding increases and reduces transplanting
	Early planting (KS)	Shifting of planting dates in Malawi (Pangapanga *et al.*, 2012)
	Home/kitchen gardens (KS, WS, GESI)	Women need less water (Carr and Thompson, 2013)
	Cover crop methods (ES, WS)	Significant changes by gender for cover-cropping in Kenya (Bernier *et al.*, 2015)
	Leaf-colour chart (NS)	
	Canopy sensors (NS)	
	Green manuring (CS, NS)	Minimal or no participation of women using GM was observed (Murray *et al.*, 2016; Khatri-Chhetri *et al.*, 2019)
Weed management	Cono-weeder (ES, GESI)	Reduction in labour hours and female-friendly technology

Continued

Table 7.1. Continued.

Categories of CSA	Key climate-smart interventions[i]	Examples of gender-differentiated revelation to CSA
Water management	Management of water-control structures (drainage management) (WS) Irrigation: drip irrigation, irrigation through solar pumps, rainwater harvesting (ES, WS, GESI) Furrow-irrigated bed planting (WS) Plastic ponds, community ponds (WS, GESI)	Reduction in water collection time for women and girls Drip irrigation in Nepal (Upadhyay, 2004) Rainwater harvesting (Carr and Thompson, 2013)
Livestock management (fodder collection and milking)	Fodder cultivation and management (fodder bank, improved varieties, silage/hay preparation) (CS, KS) Weather-friendly housing for livestock such as improved cattle sheds (WS, GESI) Livestock manure management (CS, NS) Biogas for domestic energy (ES)	Improved milk production during weather-stress conditions Better livestock management leading to secured income especially in cases of crop loss, reduced labour for livestock-related activities Increased nutrient supply for crop cultivation More women than men adopt feed management in Uganda and Kenya; compared to women, men adopt more tolerant livestock in Senegal and Kenya; mixed-result for manure management (Twyman *et al.*, 2014)
Aquaculture and capture fisheries management	Better water management with improving oxygen (WS, CS) Better feeds (CS, NS) Selective breeding/genetic improvements (KS) Integrated multitrophic aquaculture (CS) Adjusting harvest and market schedules (KS) Improved smoking kiln (CS, GESI)	No gender-differentiated research found specific to CSA in fisheries Design of gender-sensitive FAO-Thiaroye fish processing (smoking kiln) technique is used in many countries (Côte d'Ivoire, Ghana, Senegal, Tanzania, and Togo) (World Bank, FAO, IFAD, 2015)
Energy management	Energy-efficient cooking stove (CS, ES, GESI) Solar-, wind- or bioenergy-operated water pumps (WS, ES, CS)	Energy stove more attractive to women (Ragasa, 2012; Bernier *et al.*, 2013)

[i] Figures in brackets are forms of 'smartness' in CSA – WS (water-smart), CS (carbon-smart: reduces GHG), KS (knowledge-smart), ES (energy-smart), NS (nutrient-smart), GESI (gender equality and social inclusion-smart)

to be accompanied by additional research and programmes which ensure that women's labour is remunerated or the benefits are shared (ibid.).

There is limited technology for livestock, forestry and fishery, and limited research on their adoption, either generally or with a gender dimension, despite the fact that fisheries (both capture and aquaculture) and livestock activities are highly gendered. The commonly adopted practices in both sectors are improved breeding and improved feed or fodder, the purchase of which is often done by men.

Despite these biases, 'greater, differentiated, but equal access to CSA tools and techniques as well as climate services could potentially change how resources are used and improve the balance of gendered roles in agriculture' (Perch and Byrd, 2015). Influencing people or state agents at implementation stages may have a positive impact. For example, the government of Bangladesh has involved the private sector in insurance schemes for gender-responsive CSA, due to the private sector's experience with such schemes (World Bank, FAO and IFAD, 2015). Akter *et al.* (2016) have studied the preferences of weather-index insurance (WII) by 433 male and female farmers living in Bangladesh. The study found that 55% of the respondents chose traditional insurance over WII. The gender-differentiated results were: female respondents are more likely to be insurance-averse and less likely to opt for the full and partial return schemes; and owing to low financial literacy, female farmers found the conditions of WII schemes complicated.

FAO (2017) highlights the following as some of the principles for ensuring that CSA planning and implementation is gender-responsive and supportive of the poorest and most vulnerable, both to enable an effective CSA approach and to advance related Sustainable Development Goals (SDGs) (Box 7.2).

Like the FAO's suggestions, Mehar (2020) has discussed cautions and concerns for the gender dimension in CSA uptake (Box 7.3).

Box 7.2. Principles for ensuring gender-inclusive CSA. (FAO, 2017)

- Enable equal access by women, poor and marginalized food producers to the productive resources and information required to implement CSA, including land, extension services, ICTs, seed stock, etc.
- Remove financing barriers to allow women and economically marginalized food producers to invest in CSA. These barriers often relate to access to loans, savings opportunities and insurance.
- Promote opportunities for women food producers to participate and advance in sustainable value chains, e.g. through access to marketing support and training in the production of value-added products.
- Ensure small-scale food producers' and women's participation in planning, policy and budget processes, especially through subnational planning processes.
- Promote climate-smart agriculture through social protection systems. For example, social assistance programmes that help the poorest and most vulnerable to meet basic needs, while also providing training in CSA.
- Also, it is important to understand the different barriers that women and men face in adopting adaptive practices and design specific interventions that take these into account.

Box 7.3. Cautions and concerns for the gender dimension in CSA uptake. (Mehar, 2020)

Gender mainstreaming. Attention needed for inclusion in planning and policy. There is increasing evidence of gender mainstreaming in CSA projects. However, there is lack of attention in designing gender-responsive technologies. Limited budget for gender consideration and often gender is considered in implementation and analysis as 'add gender and stir' approach.
Gender analysis – slip-ups. In gender research, often the research slips to gender-women, with women as the victim. The other conceptual mistake made is considering all women and men are the same.
CSA practices – gender dichotomy. The majority of the CSA-designed technologies overlook financial resources including information and power limitations of women compared to men.
Integrating intersectional understandings in CSA. In order to address these constraints and enable women to benefit from use of CSA practices, more gender-transformative approaches are needed, such as involving men and other family members, and using participatory action research and social messaging around gender.

7.4 Summary

This chapter has presented conceptual underpinnings of climate-smart agriculture, a concept introduced in 2010 to achieve adaptation, mitigation and food security simultaneously. The chapter provides insights for adoption and impact of CSA interventions on the roles of men and women, with examples. For successful CSA implementation, differentiated need, role, impact and indirect effect of particular CSA on different genders must be taken into account in planning and implementation processes.

Note

[1] There are two generic terms related to intensification – 'agricultural intensification' and 'sustainable intensification' or 'sustainable agricultural intensification'. The former strongly relies on the use of external inputs such as fertilizers and plant protection chemicals like pesticides, and weedicides/herbicides. The latter, which is a rather 'recent' concept, strongly focuses on reducing adverse environmental impacts (Collin-Sowah, Institut für Agrarökonomie, Christian-Albrechts-Universität zu Kiel, Kiel, personal communication).

References

Aguilar, L. (2010) Establishing the linkages between gender and climate change adaptation and mitigation. In: Dankelman, I. (eds) *Gender and Climate Change: An Introduction*. Earthscan, London, pp. 173–193.

Akter, S., Krupnik, T.J., Rossi, F. and Khanam, F. (2016) The influence of gender and product design on farmers' preferences for weather-indexed crop insurance. *Global Environmental Change* 38, 217–229.

Aune, J.B., Asrat, R., Tekelehaimanot, D.A. and Bune, B.T. (2006) Zero tillage or reduced tillage: the key to intensification of the crop–livestock system in Ethiopia. In: Pender, J., Place,

F. and Ehui, S. (eds) *Strategies for Sustainable Land Management in the East African Highlands.* International Food Policy Research Institute, Washington, DC.

Below, T., Mutabazi, K., Kirschke, D., Franke, C., Sieber, S. *et al.* (2012) Can farmers' adaptation to climate change be explained by socio-economic household-level variables? *Global Environmental Change: Human and Policy Dimensions* 22, 223–235.

Bernier, Q., Franks, P., Kristjanson, P., Neufeldt, H., Otzelberger, A. *et al.* (2013) *Addressing Gender in Climate-Smart Smallholder Agriculture* (ICRAF Policy Brief No. 14). World Agroforestry Centre (ICRAF), Nairobi.

Bernier, Q., Meinzen-Dick, R.S., Kristjanson, P., Haglund, E., Kovarik, C., Bryan, E. and Ringler, C. (2015) *Gender and Institutional Aspects of Climate Smart Agricultural Practices: Evidence from Kenya.* CCAFS Working Paper No. 79, Copenhagen. CGIAR Research Program on Climate Change, Agriculture and Food Security (CCAFS).

Carr, E.R. and Thompson, M.C. (2013) *Gender and Climate Change Adaptation in Agrarian Settings.* Report prepared for the United States Agency for International Development.

Collins-Sowah, P.A. (2018) *Theoretical Conception of Climate Smart Agriculture.* Working Papers of Agricultural Policy, No. WP2018-02, University of Kiel, Department of Agricultural Economics, Chair of Agricultural Policy, Kiel, Germany.

Deressa, T., Hassan, R., Ringler, C., Alemu, T. and Yesuf, M. (2009) Determinants of farmers' choice of adaptation methods to climate change in the Nile Basin of Ethiopia. *Global Environmental Change: Human and Policy Dimensions* 19, 248–255.

FAO (2010) *'Climate-smart' agriculture: policies, practices and financing for food security, adaptation and mitigation.* Food and Agriculture Organization, Rome.

FAO (2013) *Project to Assist ERRA and its Partners in Restoring Livelihoods in Earthquake Affected Areas of Pakistan.* Food and Agriculture Organization of the United Nations, Rome. Available at: http://www.fao.org/emergencies/fao-in-action/projects/detail/en/c/180313/ (accessed 17 March 2022).

FAO (2017) *Climate Smart Agriculture Sourcebook.* Food and Agriculture Organization of the United Nations (FAO), Rome. Available at: http://www.fao.org/climate-smart-agriculture-sourcebook/en/ (accessed 23 March 2022).

Fisher, M.G., Warner, R.L. and Masters, W.A. (2000) Gender and agricultural change: crop–livestock integration in Senegal. *Society and Natural Resources* 13(3), 203–222.

Gebre, G.G., Soda, H., Rahut, D.B., Amekawa, Y., Normura, H. (2019) Gender differences in the adoption of agricultural technology: the case of improved maize varieties in southern Ethiopia. *Women's Studies International Forum* 76, 102264.

Gilligan, D.O., McNiven, S., Kumar, N., Meenakshi, J.V. and Quisumbing, A. (2013) *Who Decides to Grow Orange Sweet Potatoes? Bargaining Power and Adoption of Biofortified Crops in Uganda.* GAAP Note. International Food Policy Research Institute, Washington, DC.

Halbrendt, J., Kimura, A.H., Gray, S.A., Radovich, T., Reed, B. *et al.* (2014) Implications of conservation agriculture for men's and women's workloads among marginalized farmers in the central middle hills of Nepal. *Mountain Research and Development* 34, 214–222.

Khatri-Chhetri, A., Regmi, P., Chanana, N. and Aggarwal, P.K. (2019) Potential of climate-smart agriculture in reducing women farmers' drudgery in high climatic risk areas. *Climatic Change* 154(1), 29–42.

Kinkingninhoun-Mêdagbé, F.M., Diagne, A., Simtowe, F., Agboh-Noameshie, A.R. and Adégbola, P.Y. (2008) Gender discrimination and its impact on income, productivity, and technical efficiency: evidence from Benin. *Agriculture and Human Values* 27(1), 57–69.

Lipper, L., Thornton, P., Campbell, B., Baedeker, T., Braimoh, A. *et al.* (2014) Climate-smart agriculture for food security. *Nature Climate Change* 4, 1068–1072.

Low, J.W., Mwanga, R.O.M., Andrade, M.I., Carey, E. and Ball, A.M. (2017) Tackling vitamin A deficiency with biofortified sweetpotato in sub-Saharan Africa. *Global Food Security* 14, 23–30.

Magnan, N., Spielman, D., Gulati, K. and Lybbert, T. (2015) Information networks among women and men and the demand for an agricultural technology in India. IFPRI Discussion Paper 01411.

Mehar, M. (2016) Role of gender differentiated risk and time preferences in rice varietal selection in eastern India. PhD thesis, Department of Economics, School of Social Science, Indira Gandhi National Open University, India. Available at: https://shodhganga.inflibnet.ac.in/handle/10603/212284 (accessed 8 April 2022).

Mehar, M. (2020) Gender dimension in climate smart agriculture technology uptake. In: Sacs, C., Sexsmith, K. and Jensen, L. (eds) *Gender and Agriculture Handbook*. Routledge, Abingdon, UK. Available at: https://www.taylorfrancis.com/books/e/9780429199752/chapters/10.4324/9780429199752-18 (accessed 17 March 2022).

Mehar, M., Mittal, S. and Prasad, N. (2016) Farmers' coping strategies for climate shock: Is it differentiated by gender? *Journal of Rural Studies* 44, 123–131.

Mehar, M., Yamano, T. and Panda, A. (2017) Role of gender, risk and time preference in influencing rice varietal selection decision in eastern India. *Asian Journal of Agriculture and Development* 14(1), 18–35.

Molua, E.L. (2012) Gendered response and risk-coping capacity to climate variability for sustained food security in northern Cameroon. *International Journal of Climate Change Strategies and Management* 4, 277–307.

Murray, U., Gebremedhin, Z., Brychkova, G. and Spillane, C. (2016) Smallholder farmers and climate smart agriculture technology and labor-productivity constraints amongst women smallholders in Malawi. *Gender Technology and Development* 20(2), 117–148.

Mutenje, M., Kankwamba, H., Mangisonib, J. and Kassie, M. (2016) Agricultural innovations and food security in Malawi: gender dynamics, institutions and market implications. *Technological Forecasting and Social Change* 103, 240–248.

Naab, J.B. and Koranteng, H. (2012) Using a gender lens to explore farmers' adaptation options in the face of climate change: results of a pilot study in Ghana (CCAFS Working Paper No. 17). CGIAR Program on Climate Change, Agriculture and Food Security (CCAFS). www.ccafs.cgiar.org.

Ng'ang'a, S.K. (2021) Is investment in Climate-Smart-Agricultural practices the option for the future? Cost and benefit analysis evidence from Ghana. *Heliyon* 7(4), E06653.

Nyasimi, M., Radeny, M., Kimeli, P., Mungai, C., Sayula, G. *et al.* (2016) *Uptake and Dissemination Pathways for Climate-Smart Agriculture Technologies and Practices in Lushoto, Tanzania. CCAFS Working Paper no. 173*. CGIAR Research Program on Climate Change, Agriculture and Food Security (CCAFS), Copenhagen.

Pangapanga, P., Jumbe, C., Kanyanda, S. and Thangalimodzi, L. (2012) Unravelling strategic choices towards droughts' and floods' adaptation in Southern Malawi. *International Journal of Disaster Risk Reduction* 2, 57–66.

Paris, T. and Chi, T.T.N. (2005) The impact of row seeder technology on women labour: a case study of the Mekong Delta, Vietnam. *Gender, Technology and Development* 9(2), 158–184.

Paris, T.R., Singh, A., Cueno, A.D. and Singh, V.N. (2008) Assessing the impact of participatory research in rice breeding on women farmers: a case study in eastern Uttar Pradesh, India. *Experimental Agriculture* 44(1), 97–112.

Patle, G.T., Kuamr, M. and Khanna, M. (2019) Climate-smart water technologies for sustainable agriculture: a review. *Journal of Water and Climate Change* 11(4), 1455–1456.

Perch, L. and Byrd, R. (2015) *Gender in the CSA Discourse: Making the Case for Gender-smartness*. Working Paper series RIO no. 3.

Pircher, T., Almekinders, C.J.M. and Kamanga, B.C.G. (2013) Participatory trials and farmers' social realities: understanding the adoption of legume technologies in a Malawian farmer community. *International Journal of Agricultural Sustainability* 11, 252–263.

Ragasa, C. (2012) Gender and institutional dimensions of agricultural technology adoption: a review of literature and synthesis of 35 case studies. In: International Association of Agricultural Economists (IAAE) Triennial Conference, Fozdo Iguaçu, Brazil, 18–24 August. DOI: 10.22004/ag.econ.126747.

Sain, G., Loboguerrero, A.M. and Corner-Dolloff, C. (2017) Costs and benefits of climate-smart agriculture: the case of Dry Corridor in Guatemala. *Agricultural Systems* 151, 163–173.

Schroth, O., Pond, E. and Sheppard, S.R.J. (2015) Evaluating presentation formats of local climate change in community planning with regard to process and outcomes. *Landscape and Urban Planning* 142, 147–158.

Suckall, N., Tompkins, E. and Stringer, L. (2014) Identifying trade-offs between adaptation, mitigation and development in community responses to climate and socio-economic stresses: evidence from Zanzibar, Tanzania. *Applied Geography* 46, 111–121.

Tsephel, S., Takama, T., Lambe, F. and Johnson, F.X. (2010) Why perfect stoves are not always chosen: a new approach for understanding stove and fuel choice at the household level. *Boiling Point* 57, 6–8.

Twyman, J., Green, M., Bernier, Q., Kristjanson, P., Russo, S. *et al.* (2014) Gender and climate change perceptions, adaptation strategies, and information needs: preliminary results from four sites in Africa. CCAFS Working Paper no. 83. CGIAR Research Program on Climate Change, Agriculture and Food Security (CCAFS), Copenhagen.

Upadhyay, B. (2004) Gender aspects of smallholder irrigation technology: insights from Nepal. *Journal of Applied Irrigation Science* 39(2), 315–327.

Vandercasteelen, J., Dereje, M., Minten, B. and Taffesse, A.S. (2018) Labour, profitability and gender impacts of adopting row planting in Ethiopia. *European Review of Agricultural Economics* 45, 471–503.

Williams, P.A., Ng'ang'a, S.K., Crespo, O. and Abu, M. (2020) Cost and benefit analysis of adopting climate adaptation practices among smallholders: the case of five selected practices in Ghana. *Climate Services* 20, 100198.

World Bank Group, FAO and IFAD (2015) *Gender in climate-smart agriculture. Module 18 for Gender in Agriculture Sourcebook*. World Bank, Food and Agriculture Organization of the United Nations, and the International Fund for Agricultural Development. Available at: https://openknowledge.worldbank.org/handle/10986/22983 (accessed 10 April 2022).

Yokouchi, T. and Saito, K. (2016) Factors affecting farmers' adoption of NERICA upland rice varieties: the case of a seed producing village in central Benin. *Food Security* 8, 197–209.

8 Gender-differentiated Responses to Climate Risk: Conceptual Literature Review

Abstract

A vast array of literature has explored climate change and climate-resilient strategies with regard to gender-differentiated impacts and responses. This chapter discusses the literature on the six themes that emerged from a conceptual literature review method and uses illustrative case studies to support the arguments under discussion. The six themes are: research context, adaptation strategies, participation and empowerment, gender-inclusive climate policies, gender identity and intersectionality, and information access. The main findings show that the vast amount of available literature is skewed to a few research issues and geographical areas and has paid less attention to analysing gender beyond the men versus women dichotomy to power and social relations.

8.1 Introduction

An enormous body of literature now exists about climate-resilient strategies (CRS) in an agricultural setting in general and with a gender dimension. However, the quality and research issues vary across region, type of agricultural intervention, etc. For example, Call and Sellers (2019) found that fewer articles focused on forestry and capture fisheries and aquaculture as compared to agriculture, using a systematic literature review. On this pretext, we are curious to learn about the typology of available literature in terms of themes and topics covered, methods used, the conceptual connection of ideas to reality and the ways in which, and the degree to which, gender is addressed.

A conceptual literature review is used to identify and present gender and adaptation work. Petticrew and Roberts (2006) defined the conceptual review or conceptual synthesis method as 'a review that aims to synthesize areas of conceptual knowledge that can contribute to a better understanding of these issues' (p. 39). The purpose of each conceptual synthesis is not to provide an exhaustive

 Climate Change, Adaptation and Gender: Policy, Practice and Methodological Underpinnings (M. Mehar and N. Prasad)
DOI: 10.1079/9781789249910.0008

search and review of all the literature published but to identify the key themes and present the connection-driven (identification of relationships between ideas and practice as well as theory and reality) inferences (Fig. 8.1). The literature used here is all the relevant literature (and cross-references) the authors encountered since starting the book.

The six themes constructed to present the conceptual synthesis are:

1. Research context: In which sectors and which areas is the research focused?
2. CRS strategies: What strategies are available and adopted?
3. Participation and empowerment: How and to what degree is gender addressed?
4. Gender-inclusive climate policies: To what extent are gender issues explored and incorporated into climate policies?
5. Gender identity and intersectionality: Adaptation with the lens of intersectionality (age, class, ethnicity, etc. of respective genders) and by integrating the gender dimension in regard to who in the family makes the decisions and who takes actions.
6. Information access: Who, how and from where should we access CRS-related information, and what should we access?

Each of the themes is discussed with relevant literature and an illustrative case study (published after 2015) except for the last two themes. An illustrative case study is one type of case study, primarily descriptive and utilizing one or two instances of the event (article in our case) to depict the investigation. During the search, while exploring an article about these themes, we found at least one interesting/self-explanatory paper for each, which is used as an illustrative case study. In the last two themes, we provide summary tables of relevant references with examples.

Fig. 8.1. Conceptual synthesis of themes for gender, agriculture and climate-resilient strategies nexus. (From: the author)

8.2 Research Context

As mentioned before, there is a vast literature discussing the nexus of gender–agriculture–climate CRS actions. However, most of the studies are concentrated on crops, with less focus on forestry, fishery and livestock. Additionally, they are quite concentrated in a few regions across the world.

Bunce and Ford (2015), in their literature review of 123 documents, globally, found that significant geographic disparities exist in gender engagement, with the majority of studies from sub-Saharan Africa (40%) and Asia (20%). The authors found that almost half of the studies from Asia were from Bangladesh. In another literature review study, by Gumucio *et al.* (2020), out of 43 identified review studies, 27 were based in sub-Saharan Africa (with the majority from Uganda and Tanzania) and eight were from South and Southeast Asia. The authors also found that Oceania and north-east Africa studies are minimal and discussed not exclusively but in multi-country studies. Price (2018), who reviewed women-initiated or women-led measures for climate solutions, both global and in the context of South Asia, found that the majority of studies were from a few countries. She found most of the examples from India, followed by Bangladesh and Nepal, with less research in other South Asian countries (for example, no work was found for Afghanistan). Also, the examples of India, Bangladesh and Nepal are mostly from the Himalaya region. The author also mentioned that most of the studies are dominated by agricultural-resilient strategies followed by forestry movements.

It is important to keep in mind the research context while comparing outcomes of CRS.[1] The success of some strategy at one place may not be true or feasible in another context due to differences in agroecological conditions as well as people's attitude towards acceptance of the strategy. For example, Paris and Pingali (1996) found that the introduction of a labour-saving technology (mechanical thresher) reduced labour in the Philippines for both men and women, in contrast to in Bangladesh where it reduced the labour and thus labour-income of poor and landless women. The reason may be that in Bangladesh women, due to cultural restrictions, were not able to undertake alternative employment, whereas in the Philippines there is no such restriction.

For this theme, we consider Call and Sellers (2019) to be an appropriate illustrative case study (Box 8.1).

8.3 Which Climate-resilient Strategy(ies)?

Farmers have adopted various multiple coping strategies pertaining to particular climate shocks and their access to different resources, as well as their attributes. Investigations into coping and adaptation are often differentiated between risk-management approaches focusing on hazard-coping strategies, and adaptation considering the impacts of climate change (Agrawal, 2008). For example, selling assets or migration to other places for work are common strategies adopted by households to cope with drought or flood, whereas they may adopt crop-management practices like crop rotation or drought-resistant varieties as an *ex-ante* means to adapt to an increasing trend of climate shocks.

Box 8.1. Call and Sellers (2019).

Title: Systematic literature review of how gender vulnerability shapes the adoption and impact of sustainable livelihood interventions in an era of global climate change.
Objective: The review aims to provide a current understanding of research concerning gendered participation, adoption and the impact of agricultural technologies.
Method/design: Systematic literature review of peer-reviewed English-language articles published from January 2005 to September 2018, searched in Scopus and Web of Knowledge databases.
Review results: 131 relevant articles based on the search criteria were identified. The key inferences are:

- The majority of the articles focused on agricultural interventions (43%) followed by forestry (34%) and capture fisheries and aquaculture (24%). Most focused on adoption or participation in interventions rather than outcomes.
- The majority of the articles were quantitative-based (56%) followed by qualitative (24%) and mixed-method (19%). There was a strong geographic bias towards South Asia (Fig. 8.2).

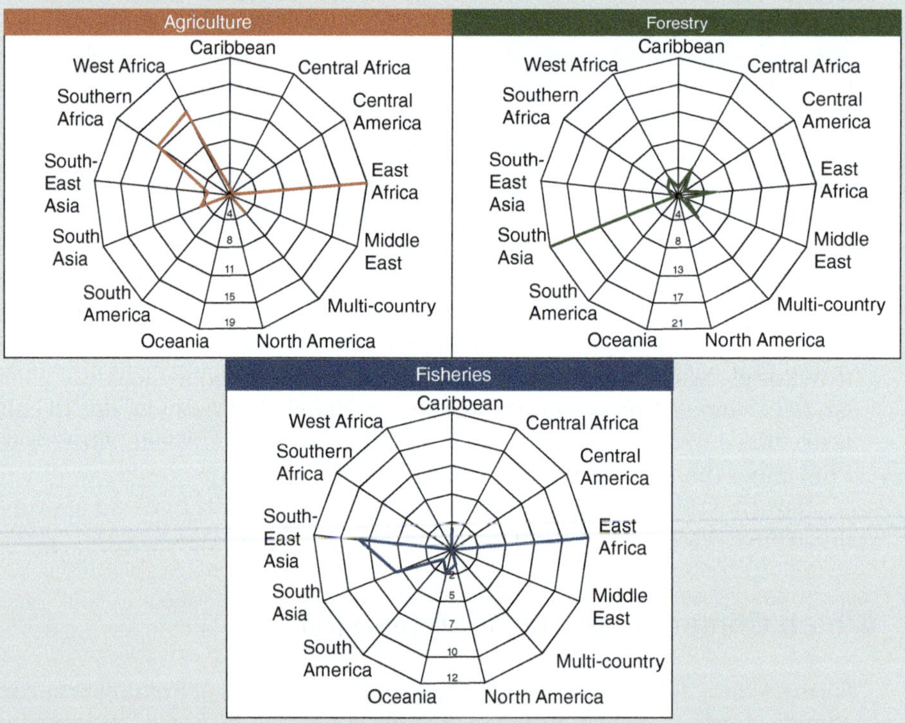

Fig. 8.2. Regional distribution of research articles by type of natural resource-based livelihood intervention.

- In general, women are less likely than men to engage with sustainable livelihood interventions. Though regional disparities were found in studies from Kenya and Benin, gender did not significantly affect the adoption of management strategies or

Continued

Box 8.1. Continued.

- inputs. However, in studies from other African countries (Nigeria, Malawi, Haiti), women were more likely to adopt agricultural technologies. Studies from Bangladesh and Malawi also report an increase in women's household decision-making powers and income as well as improving children's nutritional outcomes.
- Very few studies explored the intersectional impacts of initiatives (for instance, stratifying women's and men's adoption decisions by social class, education, ethnicity, etc.)

In the case of loss of crops and animals during and after floods and droughts, it is very difficult for a household of low resilience to sustain these losses. In such cases, both men and women have to work harder to save their crops and animals during and after disasters; other household members take part in planting more crops and raising animals, but women have increased workloads due to the need to replant rice fields and subsidiary crops (Oxfam and UNVN, 2009; Trung, 2013). Studies have found that in such situations, male farmers usually opt for off-farm activities, whereas female members reduce their food intake and expenditure on education and health (Mehar *et al.*, 2016). There are also examples where females have shifted their livelihood options because of climate change. For example, women in the Northern Territory are known to shift to aquaculture (Petheram *et al.*, 2013).

There are very few studies that have explored mitigation strategies in general and regarding research on gender and mitigation as compared to research on adaptation and impacts.

Research focusing on regions in Africa and South Asia has demonstrated that women are less likely to adopt adaptation strategies (Huyer and Partey, 2020) than men, except for a few studies where females are more likely to adopt them (Nhemachena and Hassan, 2008). Price (2018), in her literature review, found that adaptation examples are very site-specific; however, common strategies throughout rural areas in South Asia include out-migration (by men), agricultural intensification, and livelihood diversification. Chindarkar (2012) has proposed a framework to operationalize and examine vulnerability to climate change-induced migration and its gender dimensions.

Adzawla *et al.* (2019) found that, overall, both males and females in Ghana adopt strategies which include changing planting dates, row planting, planting early-maturing and drought-tolerant seed varieties, mixed farming, intercropping and refiling of farm plots. The author also found higher adoption levels by male farmers for zero tillage and intercropping. In another study in Ghana, by Wrigley-Asante *et al.* (2017), male farmers were found to cultivate drought-resistant crop varieties while female farmers engage in the cultivation of traditional crops and mixed cropping. Ashwill *et al.* (2011) found that women in Bolivia focus more on innovative approaches such as planting new crop varieties and seeking alternative water supplies, whereas men prioritize interventions such as irrigation.

Bayard *et al.* (2007) discovered that male farmers were more responsive to the adaptation of environmental degradation by planting alley crops in Haiti. Hassan and Nhemachena (2008) also indicated that farmers shifted away from monocropping and irrigated as a way of adapting to climate change during changes in temperatures. Nation (2010) documents that women in Senegal became more dependent on men after the introduction of irrigation technologies which were primarily controlled by men.

There are very few studies that did not find any gender differences in the adoption of examined adaptation strategies in respective studies. For example, Muriithi *et al.* (2018) did not find any gender differences in adoption of maize-grain legume intercropping, crop rotation, fertilizer use and improved maize seeds. However, the authors found that jointly managed (by both men and women) plots were more likely to adopt strategies such as crop rotation and soil and water conservation, as compared with male-managed plots.

In addition to adoption, there are a few studies which explore the trade-offs associated with technology. For example, the lack of herbicides while using zero tillage (ZT) may increase weed infestation and thus the demand for women labour (Tsige *et al.*, 2020). Such examples are specified in the contested climate-smart agriculture and gender discussion in Chapter 7.

Mehar *et al.*'s (2016) article is being considered as a case study as it is the only study to assess gender of decision makers and action takers in respect of the selected strategies (both off-farm and on-farm) (Box 8.2).

8.4 Participation and Empowerment

Recognizing and considering the gender aspect in the design, implementation and adoption of adaptation strategies and programmes is one aspect. The other aspect is the degree to which gender-sensitivity is applied in all aspects for the success of the programme. Most of the studies engaged gender during the survey stage rather than in planning the research. Bunce and Ford (2015) is being used as an illustrative case study (Box 8.3).

In the past few years, there have been increased efforts to apply a degree of gender sensitivity in programmes. Section 2 of Chapter 3 has more details about this. Most of the studies are participatory, giving evidence for stress-tolerant varieties of different crops, livestock (animals and their feed) as well as fish (species and feed); for example, climate-resilient sweet potato seed and cassava seed. Failure to engage gender in these studies could result in maladaptation, which may have uneven impacts across the gender spectrum (Bunce and Ford, 2015). For example, a case study on solar home systems in Bangladesh, where women were nominated for participation, as men work outside, on the sea and other activities, found that women learning new skills and participating in the community, and earning money, did not necessarily challenge gender inequality (Wong, 2009). Women's decisions were often challenged by men after they returned from fishing. The case study also found an increased burden of working hours for women due to the increased availability of light, whereas men use that same time for leisure, e.g. television (ibid.). Thus it is important to understand the gender norms and integrate them into the planning for gender-effective climate responses. It is not only important to ensure access of both men *and* women to new technologies but also to understand gender risks and opportunities attached to them.

Box 8.2. Case study for decision making using Mehar *et al.* (2016).

Title: An empirical approach to understand how farmers select coping strategies to manage climate shock
Objective: The paper tests the hypothesis that male and female farmers have different coping strategies and that different socio-economic factors influence decision-making capability.
Method/design: The paper uses baseline data of the CCAFS project collected in 2013 using a multi-stage sampling method in the Vaishali district of Bihar state, India. The farmers in the households surveyed reported that they adopt multiple coping strategies after the occurrence of climate shocks like flood and drought. The paper tests the hypothesis using a multivariate probit model, which allows for the possible correlation in the choice to use six coping strategies simultaneously.
Review results: The key results are:

- The study observes that there exist gender differences in access to resources.
- Around 15% of surveyed farmers reported that they had not undertaken any coping strategy after a stressful event like drought or flood. The rest of the households adopted at least one coping strategy.
- The study found that farmers adopt multiple strategies to cope with risk.
- The results suggest that the gender of the decision maker plays a significant role in the selection of coping strategies from the major categories. However, the gender of the action taker is only significant for the coping strategy of crop rotation.
- The gender of a farmer is found to have a significant role in selecting strategies. For instance, crop rotation was selected by 60% of male farmers. However, considering simultaneous decisions regarding coping strategies, the authors found that there is no significant difference between male and female decision makers when compared with the base category of a collective decision by all family members.
- The most common strategy adopted by males and females is cutting down the intake of food, if the climate shock impact is drastic.
- In contrast to gender role in solo decisions, where gender significantly influences the decision, in simultaneous decision making, the household characteristics are more significant in influencing the decision.

Exposure to extension and training has a positive influence on adapting agriculture-specific coping strategies. However, female farmers have little access to extension services and therefore programmes and policies aimed at helping families cope with climate change need to address this inequity.

Though gender integration is important, social norms may challenge the dissemination process. For example, Mudege *et al.* (2017) explored gender conceptions about training for nutritious climate-resilient orange-fleshed sweet potato in Malawi. In this programme, men mainly received production-orientated training, whereas women were the majority of participants attending training in processing and nutrition. In the focus group discussion, women expressed an interest in training, but when men were questioned (who are farmers as well as extension workers), they mentioned that women were lazy or simply not interested in attending the sessions.

Box 8.3. Bunce and Ford (2015).

Title: How is adaptation, resilience and vulnerability research engaging with gender?
Objective: Assess how climate change adaptation, resilience and vulnerability (ARV) research is engaged with gender in peer-reviewed literature.
Method/design: Systematic literature review of peer-reviewed articles published between 2006 and 2014 in Scopus and Web of Knowledge portals. A conceptual model capturing attributes of gender 'engagement' was developed.

Table 8.1. Gender mainstreaming attributes and exemplary questions.

Attributes and components of engagement	Questions/indicators
Gender sensitivity	Is there explicit reorganization of the different needs and experiences by gender?
	Is gender-sensitive language used?
Gender-responsiveness	Are the research findings presented in a gender-disaggregated manner?
	Do progress indicators measure the different impacts experienced by each gender?
Gender-transformativeness	Does the research critically analyse social values, organization practices and goals?

Review results: After the screening, 123 documents were used for examining gender in the context of climate change ARV. The key results were:

- An increasing trend of ARV studies both with a gender focus and with the level of gender engagement has been observed. Though the studies focus on those outside the gender binary, in most studies a predominant focus is on women as the understudied gender.
- In the reviewed studies, women are either being described as a vulnerable population or viewed as active participants in responding to climate change impacts.
- Adaptation-focused studies have higher levels of engagement than vulnerability work, with resilience studies having a limited gender focus.
- Overall, the levels of gender-sensitive and gender-responsive indicators were higher than for gender-transformativeness, reflecting societal norms.
- Significant geographic disparities exist in gender engagement, with the majority of studies from sub-Saharan Africa (40%) and Asia (20%), of which almost half were from Bangladesh.

Table 8.2. Gender mainstreaming performance of ARV literature by attribute (% of the article).

	Gender sensitivity	Gender responsiveness	Gender transformativeness
Adaptation	83	85	57
Vulnerability	69	70	46
Resilience	50	50	38

Source: Table 6 of the supplementary data of the article

Most reviewed articles focus on practical gender needs in relation to climate change than on strategic need; 25% of the paper failed to discuss either.

There have been a few attempts to influence the gender norms and resultant increase in climate-smart agriculture (CSA) scaling. For example, Salvini *et al.* (2016) used a role-playing game and collected pre- and post-game information related to adoption decisions. The authors found the game had a significant impact on the adoption of CSA. There are limited published examples of this. One constraint may be that role-playing games are often contextual and often lack enough data to support an academic paper.

8.5 The Extent of Gender-inclusiveness in Climate Policies

As climate change continues its trend of losses in production and rising inequality, international agreements (refer to Box 1.1 for general climate policies and Chapter 3 for gender-specific programmes) have encouraged national governments to implement gender-responsive climate strategies. Khan and Roberts (2013) reviewed the development of policies, institutions and financing of adaptation in international agreements from 1992 to the present. The authors conclude: '...the way the treaty has been built – first as a mitigation regime with adaptation added on only later – has led to some profound problems for marrying the goals of economic development and building climate-resilient societies'. As regards the geographical trend, the number of adaptation policies and practices has increased in African and low-income countries, particularly those nations receiving adaptation funds since 2006, whereas North Africa and Central Asia have less documentation (Ford *et al.*, 2015). Gumucio and Mariana (2015) have reviewed the extent of gender inclusion in national policies related to climate change in seven countries in Latin America. Singh *et al.* (2021), assessing 28 State Action Plans on Climate Change (SAPCC) in India, found that 43% had no mention of gender, either in discussions on vulnerability or in adaptation actions. In a similar study, in Nepal, Rai *et al.* (2021) found, out of 24 climate change policy documents, that five were gender-blind and 19 were gender-aware.

National policies are often influenced by gendered social norms, are gender-biased and reinforce and reproduce masculine values. In such scenarios, the adaptation strategies implemented exacerbate the existing gender inequalities. Acosta *et al.* (2021), presenting views from national and sub-national policy actors in Uganda using semi-structured interviews, found that local actors tended to adhere to global discourses on gender inequality when devising context-specific solutions to agriculture and climate change. However, their proposed solutions did not challenge the gender norms and culture, which are major barriers for gender-equal formal policies. Ford *et al.* (2015) found limited evidence of adaptation initiatives being targeted at vulnerable populations including socio-economically disadvantaged populations, children, indigenous peoples and the elderly. The authors also recognize the need for enhancing linkages between national planning and local initiatives (ibid.).

Mersha and van Laerhoven's (2019) article examining Ethiopian climate-resilient national policies through the lens of gender is used as a case study here (Box 8.4). Another reference for a good case study is Jerneck (2018).

Box 8.4. Mersha and van Laerhoven (2019).

Title: Gender and climate policy: a discursive institutional analysis of Ethiopia's climate-resilient strategy
Objective: The study develops a twofold objective: first, identifying the extent to which gender is represented in adaptation strategies in national-level climate policies; and, second, exploring the way in which, and the extent to which, a gendered co-ordinative discourse affects policy outcomes.
Method/design: For the first objective, the authors adopt three nested levels of ideas (policies, programmes and philosophies) of the 'discursive institutionalism' (DI) approach. They link the ideas with three key questions in adaptation analysis, namely: (i) Adaptation to what and why?; (ii) Who or what adapts?; and (iii) How does adaptation occur? The second objective adopts the co-ordinative discourse (CD) approach to assess how, when, to whom and why policy ideas are conveyed.
The assessment is based on 20 semi-structured interviews with policy actors (selected using purposive and snowball sampling) in government ministries and staff members of local civic society groups and international organizations.
Review results: Overall, the results found a dominance of gender-neutral wording in the climate-resilient strategies for agriculture and forestry in Ethiopia. Inferences based on analysis of the specific questions of the two approaches are:

- DI: Adaptation to what and why? The climate policy documents were found to contain details about climate issues, avoiding mentioning the gender dimension and social aspects.
- DI: Who or what adapts? The document cites women, people with disabilities and children as the most vulnerable, but with no equivalent analysis of how and why these different groups were included in this category. This grouping not only reflects an essentialist assumption about women but is also paternalistic, as the root causes of women's vulnerability remain largely concealed.
- DI: How does adaptation occur? As the vulnerability assessment did not consider gendered power relations, institutions and other socio-economic drivers, gender-responsive adaptation options are hardly considered at this level.
- CD: How, when, to whom and why? The findings of CD assessment are a limited number of representatives in climate policymaking. Those included have limitations in terms of expertise in climate change-related discourses. An impression from the interview expounds as follows:
  - 'Regarding gender-related questions, the consultants often appeared unprepared' (verbatim record).
  - 'There is no climate change expert in our department' [Women Affairs Directorate]. 'I only took a one-day training on the issue. Thus, the knowledge gap also restrains us from having a meaningful engagement' (verbatim record).

The CD assessment also shows that gender ideas were entered as an add-on concern at a relatively late stage in framing policies. The authors have recommended the 'necessity of considering a gender agenda from the very initial stages such as the development of TOR (Terms of Reference) and the importance of having a sufficient number of capable actors to represent the gender dimension in co-ordinative discourses'.

8.6 Gender Identity and Intersectionality

Gender identity defined by power and position in the family may have a huge role in the adoption of adaptation strategy; so it is important to know who in the family makes decisions and takes action about adapting mitigating strategies. Based on culture and social norms as well as household dynamics, the person can be viewed from different positions. This is a strong influencing reason for adaptation but is not fixed in a wider context. The neoclassical approach was to interview the heads of households who have control over the family and influence decisions. Several studies to date have used this gender identity to discuss adaptation strategy decisions. However, there are other studies which have given weight to other identities like husband and wife, plot owner, decision maker as well as multiple others. Table 8.3 provides exemplary references by gender identity exploring the gender–agriculture–adaptation strategy nexus. Different identities may have different results. For example, Chirwa (2005) used multiple identities for the adoption of fertilizer and hybrid maize in Malawi. He found that no significant association exists between the gender of the plot manager and adoption rates but did find that female-headed households had lower adoption rates than their male counterparts.

Besides gender identity, the other consideration is that not all women and all men are homogenous in their respective gender. There exist multiple and fragmented heterogeneity features which differ from woman to woman and man to man. The features are socio-economic characteristics such as age, caste, wealth, etc., which influence the power dynamics and ultimately the negotiation powers of the person. For example, Carr and Thompson (2013), cited in Thompson-Hall *et al.* (2016), discuss the age factor with adaptation decisions of rain-fed crops, which influence relevant adaptation strategies. They stated that senior women are more dependent on sales of surplus rain-fed crops from the hand-irrigated garden in contrast to junior women, whereas in the same study, junior men are more reliant on the sale of a surplus rain-fed crop than senior men. Similar inferences with gender and age intersectionality are drawn by Carr *et al.* (2016) in

Table 8.3. Exemplary references by gender identity.

Gender identity	Exemplary references
Head of household	Bayard *et al.* (2007); Rioux *et al.* (2016); Assan *et al.* (2018)
Husband and wife	Ngigi *et al.* (2017); Van Aelst and Holvoet (2017)
Plot owner	Muriithi *et al.* (2018); Ndiritu *et al.* (2014)
Decision maker & action taker	Mehar *et al.* (2016); Mehar *et al.* (2017)
Adult male and adult female	Ashwill *et al.* (2011); Twyman *et al.* (2014)
Participant of training	Salvini *et al.* (2016); Mudege *et al.* (2017)
Participants/members of groups/communities or programmes	Djoudi and Brockhaus (2011); Partey *et al.* (2018)
Multiple identities	Adzawla *et al.* (2019); Fisher and Carr (2015); Tsige *et al.* (2020)
Not defined clearly	Onta and Resurrección (2011)

Senegal and Carr and Owusu-Daaku (2016) in Mali. Another example of age intersectionality could be seen in the work of Ravera *et al.* (2016) where she found that reactive strategies of livelihood security were mostly adopted by farmers older than 60; and middle-aged people, especially women, are involved in agriculture, marketing and decision making in their households. A few studies have found that the marital status of respondents influenced their behaviour for climate-smart intervention; for example Tsige *et al.* (2020). Prime livelihood strategies also make a difference in adaptive strategies. For example, women from the agricultural community invested in charcoal production, whereas women from the pastoral community increase livestock (sedentary) (Brockhaus *et al.*, 2013). Onta and Resurrección (2011) discuss caste as an intersectionality variable in the changing roles of men and women in relation to CRS.

In a study by Djoudi and Brockhaus (2011), men and women were found to have differences in preference ranked in the gender-participatory workshop which assessed vulnerability and adaptive strategies by the participants. The authors found that male participants see migration as an immediate coping strategy, and refilling the lake in the community, hence ensuring water supply, as the most important long-term strategy, to cope with recurring droughts; whereas women do not cite lake-refilling strategy, as they did not find it practically possible. Indeed, women participants found investing in education for their children to be a long-term strategy (ibid.).

Examples of references with intersectional indicators for the gender–agriculture–CRS nexus are mentioned in Table 8.4. Studies such as Djoudi *et al.* (2016), Ravera *et al.* (2016) and Thompson-Hall *et al.* (2016) have provided a review of gender adaptation with an intersectionality dimension. There is no denying that in recent years there have been studies which looked beyond the men–women binary dichotomy and discussed other dimensions. However, most of the papers discuss the gender-embedded intersectionality aspect in their introduction and conceptualization part only. The results section and conclusions in most studies are explored with the gender binary concept and not with the intersectionality dimension.

8.7 Climate-related Information Access

One persistent puzzle is that the technologies that would minimize adverse environmental effects and increase long-term yields exist but have yet to be adopted on a wide scale, and the potential constraint identified is lack of credible information

Table 8.4. Exemplary references by intersectionality indicators.

Intersectionality indicator	Reference
Age	Carr and Owusu-Daaku (2016); Ravera *et al.* (2016)
Ethnicity	Onta and Resurrección (2011)
Social relations	Thomson *et al.* (2016)
Profession	Brockhaus *et al.* (2013)
Risk perception	Mehar *et al.* (2016)

(Mittal and Mehar, 2013, 2015; Mehar, 2014). The attribution of information on farm productivity cannot be calculated directly. However, as discussed repeatedly in this book, choices and actions made for CRS are highly site-specific and farmers need timely and relevant information about the respective strategy in addition to information about the weather forecast. Jensen (2007) did the pioneering study and discussed the impact of information value on fishermen's lives in Kerala, India. It is important to note that, from a similar time, mobile tele-density,[2] globally, including rural areas, has shown a tremendous increase. Though the Jensen study has not explored information with a gender dimension, their results are male-orientated. For example, timely information via mobile phone reduces the price dispersion by linking fishermen and wholesalers, providing them with market information to sell (Jensen, 2007), in addition to saving the lives of fishermen. Since then, there have evolved many agriculture-related ICT-based models, especially mobile-based information, which started providing weather-related information to farmers and later were upgraded with other relevant information. An example of an ICT-based evolution in India can be found in Mittal *et al.* (2019).

Fig. 8.3 portrays how different forms of climate information can help farmers to make decisions. The most recent study, by Gumucio *et al.* (2020), has reviewed the existing empirical knowledge-based trends; differences in access, use, benefit from climate services, exploring promising paths for making climate services more gender-responsive. The authors found 43 peer-reviewed publications from 2000 to 2018, globally. One of these articles is Rengalaskhmi *et al.* (2018), the findings of which are documented in Gumucio *et al.* (2020) as follows:

> Men in the study site may be interested in the weather to address concerns regarding crop selection, an issue for which they tend to be the prime decision-makers; in contrast, women's concerns for the harvest and the need for tree sub-products and crop residues to feed their goats can influence their interest in weather information.

8.8 Summary

The key results of the conceptual synthesis of the gender–agriculture–CRS nexus are:

- The majority of scientific literature available focuses on agriculture with few articles on livestock, fisheries and forestry. Within agriculture, most of the articles are on crops with negligible articles on pulses, fruits and vegetables, etc.
- The studies are mostly focused on Africa and the South Asia region. Within Africa, most studies are from Uganda, Tanzania and Kenya, whereas in South and Southeast Asia the majority of studies are in India, Bangladesh, Nepal and a few in the Philippines.
- Studies are dominated by adaptation followed by climate-smart agriculture and almost negligible research on mitigation.
- Except for reviews, most of the studies are local case studies and have context-specific results and scaling suggestions.

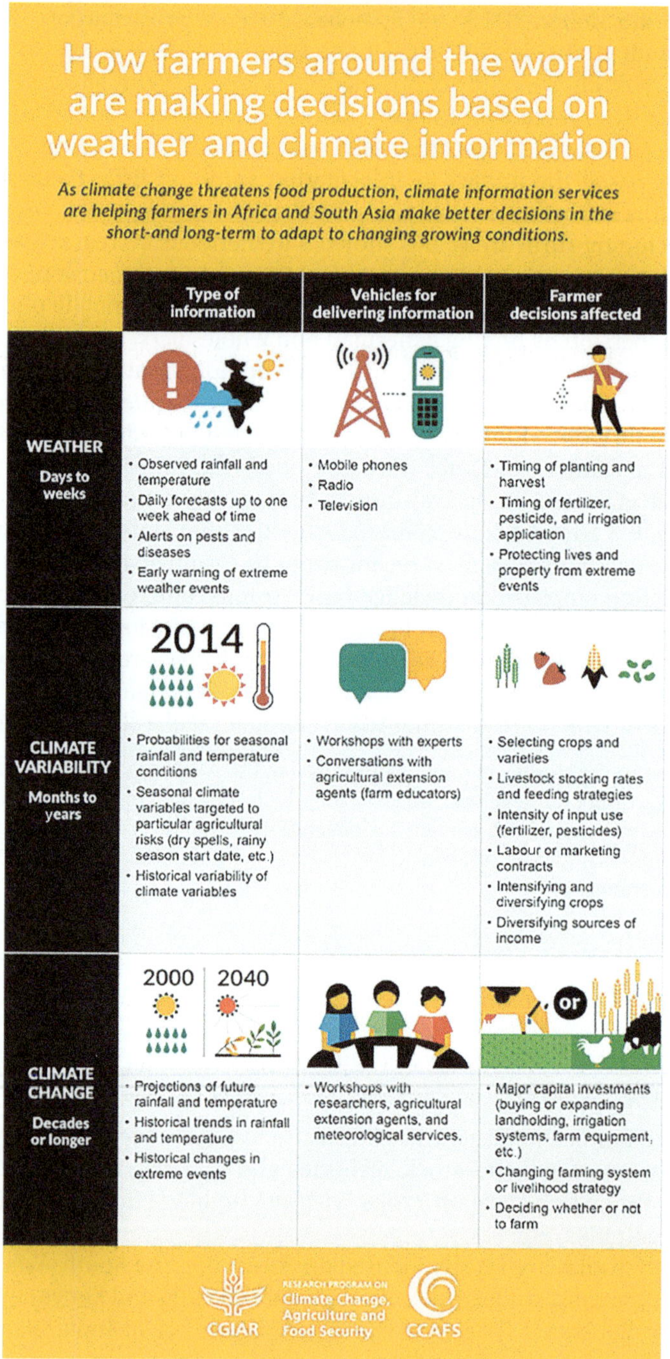

Fig. 8.3. How farmers make decisions based on weather and climate information. (From: https://ccafs.cgiar.org/sites/default/files/images/CCAFS-Climate_Servies_Infographic-BLOG-final.jpg [accessed 20 March 2022]; adapted with permission from Global Communications & Knowledge Manager, CCAFS)

Table 8.5. Examples of variables making gender differences for climate-resilient strategies.

Category	Different variables (examples)
Socio-economic characteristics	Age, household size, marital status, class
Resources	Assets (agriculture and non-agriculture)* including livestock and fishery, land entitlement, financial capital or credit facilities, natural resources, supporting labour within and outside the family, information access
Motivational	Self-efficacy, education, health status
Preferences	Technology choice, investment choices (trade-off for agriculture over education, health or diets)
Institutions	Membership or leadership position in the community, market access, social norms, variation in access to extension services

*Agriculture input includes fertilizer, seed, etc.; non-agriculture input includes jewellery, etc. The latter is also important as it can be used to raise money in shock situations.

On this pretext, it is important to note that the studies have discussed differences in the ability to adapt to and cope with changes due to different adaptive capacity and social norms of different genders. Broadly speaking, the differential impacts can be linked to: (i) access to and control over natural resources such as land, water and energy; and (ii) participation in decision making on the management of resources both within the household and at greater institutional scale. From the reviews explored in this chapter, we summarize the variables determining gender differences in Table 8.5.

Therefore, in conclusion, women play a very crucial role in climate change adaptation and mitigation, even though their contribution is often overlooked or less acknowledged. For this reason, as well as due to lack of data, there is little existing research considering the linkages between climate change and gender.

Notes

1 A specific strategy is also conducive to specific environment and agroecological conditions.
2 Mobile tele-density is the number of mobile phone users per 100 persons.

References

Acosta, M., van Wessel, M., van Bommel, S. and Feindt, P.H. (2021) Examining the promise of 'the local' for improving gender equality in agriculture and climate change adaptation. *Third World Quarterly* 42(6), 1135–1156.

Adzawla, W., Azumah, S.B., Anani, P.Y. and Donkob, S.A. (2019) Gender perspectives of climate change adaptation in two selected districts of Ghana. *Heliyon* 5: e02854.

Agrawal, A. (2008) *The Role of Local Institutions in Adaptation to Climate Change*. Social Dimensions of Climate Change, Washington, DC.

Ashwill, M., Blomqvist, M., Salinas, S. and Ugaz-Simonsen, K. (2011) *Gender Dynamics and Climate Change in Rural Bolivia*. The World Bank, Washington, DC.

Assan, E., Suvedi, M., Olabisi, L.S. and Allen, A. (2018) Coping with and adapting to climate change: a gender perspective from smallholder farming in Ghana. *Environments* 5(86), 1–19.

Bayard, B., Jolly, C.M. and Shannon, D.A. (2007) The economics of adoption and management of alley cropping in Haiti. *Journal of Environmental Management* 85, 62–70.

Brockhaus, M., Djoudi, H. and Locatelli, B. (2013) Envisioning the future and learning from the past: adapting to a changing environment in northern Mali. *Environmental Science & Policy* 25, 94–106.

Bunce, A. and Ford, J. (2015) How is adaptation, resilience, and vulnerability research engaging with gender? *Environmental Research Letters* 10, 123003.

Call, M. and Sellers, S. (2019) How does gendered vulnerability shape the adoption and impact of sustainable livelihood interventions in an era of global climate change? *Environmental Research Letters* 14, 083005.

Carr, E.R. and Owusu-Daaku, K.N. (2016) The shifting epistemologies of vulnerability in climate services for development: the case of Mali's agrometeorological advisory programme. *Area* 48, 7–17.

Carr, E.R., Fleming, G. and Kalala, T. (2016) Understanding women's needs for weather and climate information in agrarian settings: the case of Ngetou Maleck, Senegal. *Weather, Climate and Society* 8, 247–264.

Chindarkar, N. (2012) Gender and climate change-induced migration: proposing a framework for analysis. *Environmental Research Letters* 7(2), 1–7.

Chirwa, E.W. (2005) Adoption of fertilizer and hybrid seeds by smallholder maize farmers in southern Malawi. *Development Southern Africa* 22(1), 1–12.

Djoudi, H. and Brockhaus, M. (2011) Is adaptation to climate change gender neutral? Lessons from communities dependent on livestock and forests in Northern Mali. *International Forestry Review* 13(2), 123–135.

Djoudi, H., Locatelli, B., Vaast, C., Asher, K., Brockhaus, M. and Sijapati, B.B. (2016) Beyond dichotomies: gender and intersecting inequalities in climate change studies. *Ambio* 45 (Suppl. 3), S248–S262.

Fisher, M. and Carr, E.R. (2015) The influence of gendered roles and responsibilities on the adoption of technologies that mitigate drought risk: the case of drought-tolerant maize seed in eastern Uganda. *Global Environmental Change* 35, 82–92.

Ford, J.D., Berrang-Ford, L., Bunce, A., McKay, C., Irwin, M. *et al.* (2015) The status of climate change adaptation in Africa and Asia. *Regional Environmental Change* 15, 801–814.

Gumucio, T. and Mariana, T.R. (2015) Influencing gender-inclusive climate change policies in Latin America. *Journal of Gender, Agriculture and Food Security* 1(2), 42–61.

Gumucio, T., Hansen, J., Huyer, S. and van Huysen, T. (2020) Gender-responsive rural climate services: a review of the literature. *Climate and Development* 12(3), 241–254.

Hassan, R. and Nhemachena, C. (2008) Determinants of African farmers' strategies for adaptation to climate change: multinomial choice analysis. *African Journal of Agricultural and Resource Economics* 2(1), 83–104.

Huyer, S. and Partey, S. (2020) Weathering the storm or storming the norms? Moving gender equality forward in climate-resilient agriculture. *Climate Change* 158, 1–12.

Jensen, R. (2007) The digital provide: information (technology), market performance and welfare in the South Indian fisheries sector. *Quarterly Journal of Economics* 122(3), 879–924.

Jerneck, A. (2018) Taking gender seriously in climate change adaptation and sustainability science research: views from feminist debates and sub-Saharan small-scale agriculture. *Sustainability Science* 13(2), 403–416.

Khan, M.R. and Roberts, J.T. (2013) Adaptation and international climate policy. Wiley Interdisciplinary Reviews. *Climate Change* 4(3), 171–189.

Mehar, M. (2014) Gender specific impacts of climate variation in agriculture. Unpublished MPhil thesis, Department of Economics, School of Social Science, Indira Gandhi National Open University, India.

Mehar, M., Mittal, S. and Prasad, N. (2016) Farmers' coping strategies for climate shock: Is it differentiated by gender? *Journal of Rural Studies* 44, 123–131.

Mehar, M., Yamano, T. and Panda, A. (2017) Role of gender, risk and time preference in influencing rice varietal selection decision in eastern India. *Asian Journal of Agriculture and Development* 14(1), pp. 18–35.

Mersha, A.A. and van Laerhoven, F. (2019) Gender and climate policy: a discursive institutional analysis of Ethiopia's climate resilient strategy. *Regional Environmental Change* 19, 429–440.

Mittal, S. and Mehar, M. (2013) *Agricultural Information Networks, Information Needs and Risk Management Strategies: A Survey of Farmers in Indo-Gangetic Plains of India*. CIMMYT SEP Working Paper.

Mittal, S. and Mehar, M. (2015) Socio-economic factors affecting adoption of modern ICT by farmers in India: analysis using a multivariate probit model. *Journal of Agricultural Education and Extension* 22(2), 199–212.

Mittal, S., Mehar, M. and Hariharan, V. (2019) Information and communication technologies for strengthening extension services to reach the last mile in India. In: Babu, S.C. and Joshi, P.K. (eds) *Agricultural Extension Reforms in South Asia: Status, Challenges, and Policy Options*. Academic Press, UK.

Mudege, N.N., Mdege, N., Abidin, P.E. and Bhatasara, S. (2017) The role of gender norms in access to agricultural training in Chikwawa and Phalombe, Malawi. *Gender, Place & Culture* 24(12), 1689–1710.

Muriithi, B.W., Menale, K., Diiro, G. and Muricho, G. (2018) Does gender matter in the adoption of push-pull pest management and other sustainable agricultural practices? *Evidence from Western Kenya. Food Security* 10(2), 253–272.

Ndiritu, S.W., Kassie, M. and Shiferaw, B. (2014) Are there systematic gender differences in the adoption of sustainable agricultural intensification practices? Evidence from Kenya. *Food Policy* 49, 117–127.

Nation, M. (2010) Understanding women's participation in irrigated agriculture: a case study from Senegal. *Agriculture and Human Values* 27(2), 163–176.

Ngigi, M.W., Mueller, U. and Birner, R. (2017) Gender differences in climate change adaptation strategies and participation in group-based approaches: an intra-household analysis from rural Kenya. *Ecological Economics* 138, 99–108.

Nhemachena, C. and Hassan, R. (2007) *Micro-Level Analysis of Farmers' Adaptation to Climate Change in Southern Africa*. IFPRI Discussion Paper No. 00714, International Food Policy Research Institute, Washington DC.

Onta, N. and Resurrección, B.P. (2011) The role of gender and caste in climate adaptation strategies in Nepal. *Mountain Research and Development* 31(4), 351–356.

Oxfam and UNVN (2009) Responding to climate change in Vietnam: opportunities for improving gender equality. Policy discussion paper, UN Vietnam, Hanoi.

Paris, T.R. and Pingali, P.L. (1996) Do agricultural technologies help or hurt poor women? Competition and conflict in Asian agricultural resource management: issues, options, and analytical paradigms. IRRI Discussion Paper Series No. 11, pp. 237–245. International Rice Research Institute (IRRI), Los Banos, Lagun.

Partey, S.T., Dakorah, A.D., Zougmore, R.B., Ouedraogo, M., Nyasimi, M., Nikoi, G.K. and Huyer, S. (2018) Gender and climate risk management: evidence of climate information use in Ghana. *Climatic Change* 158, 61–75.

Petheram, L., Fleming, A., Stacey, N. and Perry, A. (2013) *Indigenous Women's Preferences for Climate Change Adaptation and Aquaculture Development to Build Capacity in the Northern Territory*. National Climate Change Adaptation Research Facility, Gold Coast, Australia.

Petticrew, M. and Roberts, H. (2006) *Systematic Reviews in the Social Sciences: A Practical Guide*. Blackwell Publishing, Malden, Massachusetts and Oxford, UK.

Price, R. (2018) Women-Initiated Measures to Cope with Environmental Stresses and Climate Change in South Asia. Institute of Development Studies. Helpdesk Report.

Rai, A., Ayadi, D.P., Shrestha, B. and Mishra, A. (2021) On the realities of gender inclusion in climate change policies in Nepal. *Policy Design and Practice* 4(4), 501–516.

Ravera, F., Martín-López, B., Pascual, U. and Drucker, A. (2016) The diversity of gendered adaptation strategies to climate change of Indian farmers: a feminist intersectional approach. *Ambio* 45, S335–S351.

Rengalakshmi, R., Manjula, M. and Devaraj, M. (2018) Making climate information communication gender sensitive: lessons from Tamil Nadu. *Economic & Political Weekly* LIII (17), 87–95.

Rioux, J., Gomez San Juan, M., Neely, C., Seeberg-Elverfeldt, C., Karttunen, K. *et al.* (2016) *Planning, Implementing and Evaluating Climate-Smart Agriculture in Smallholder Farming Systems: The Experience of the MICCA Pilot Projects in Kenya and the United Republic of Tanzania*. FAO, Rome.

Salvini, G., van Paasen, A., Ligtenberg, A., Carrero, G.C. and Bregt, A.K. (2016) A role-playing game as a tool to facilitate social learning and collective action towards Climate Smart Agriculture: lessons learned from Apuí, Brazil. *Environmental Science & Policy* 63, 113–121.

Singh, C., Solomon, D. and Rao, N. (2021) How does climate change adaptation policy in India consider gender? An analysis of 28 state action plans. *Climate Policy* 21(7), 958–975.

Thomson, M.C., Carr, E.R. and Pascual, U. (2016) Enhancing and expanding intersectional research for climate change adaptation in agrarian settings. *Ambio* 45(Suppl 3), 373–382.

Thompson-Hall, M., Carr, E.R. and Pascual, U. (2016) Enhancing and expanding intersectional research for climate change adaptation in agrarian settings. *Ambio* 45, 373–382.

Trung, P. (2013) Climate change and its gendered impacts on agriculture in Vietnam. *International Journal of Development and Sustainability* 2(1), 52–62.

Tsige, M., Synnevag, G., Aune, J.B. (2020) Gendered constraints for adopting climate-smart agriculture amongst smallholder Ethiopian women farmers. *Scientific African* 7, e00250.

Twyman, J., Green, M., Bernier, Q., Kristianson, P., Russo, S. *et al.* (2014) *Adaptation Actions in Africa: Evidence that Gender Matters. CCAFS Working Paper no. 83*. CGIAR Research Program on Climate Change, Agriculture and Food Security (CCAFS). Copenhagen.

Van Aelst, K. and Holvoet, N. (2017) Climate change adaptation in the Morogoro region of Tanzania: women's decision-making participation in small-scale farm households. *Climate and Development* 10(6), 495–508.

Wong, S. (2009) Climate change and sustainable technology: re-linking poverty, gender, and governance. *Gender and Development* 17(1), 95–108.

Wrigley-Asante, C., Owusu, K., Egyir, I.S. and Owiyo, T.M. (2017) Gender dimensions of climate change adaptation practices: the experiences of smallholder crop farmers in the transition zone of Ghana. *African Geographical Review* 1–14.

9 Unpacking the Heterogeneous Nature of Climate Stresses and Varying Gendered Roles Across Crop Cycles

Abstract

The cost of not including men's and women's preferences and needs in climate change strategies is enormous. Their needs and preferences are influenced by their roles and responsibilities in agriculture and household-related decisions and activities. It will not be appropriate to evaluate their needs in the overall production (or crop) cycle; their involvement and actions vary in different stages of the crop cycle. This chapter presents a conceptual framework drawn from 'livelihood diversification theory' and the 'characteristic model'. Men's and women's involvement varies not only at different crop stages but climate-resilient strategies (CRS) are also relevant to different crop stages, as they are designed based on at what stage climate shocks can occur.

9.1 Introduction

In a world free of risk and uncertainty, the decision-making functions of farmers would be greatly simplified; however, this is not the case. In crop production processes (or even in the growth cycle of livestock and fish), the effects of factors such as weather and other natural phenomena are not known until the crop (or fish) is harvested. Climate stress is *prima facie* evidence for yield reduction and an increase in social inequality, albeit climate is only one stress among many faced by the farmer. Ignoring another related aspect may lead to overlooking climatic stress importance. Stirling and Scoones (2009) stated, 'reductive–aggregative risk-assessment techniques may provide powerful responses to a narrow state of risk not applicable to less tractable conditions of uncertainty, ambiguity, and ignorance', and 'they yield enormous (if implicit) variabilities under a range of circumstances'. Adapting their four forms of incertitude (risk, uncertainty, ambiguity and ignorance), farmers' concerns in the context of climate change and CRS decisions are framed (Fig. 9.1).

 Climate Change, Adaptation and Gender: Policy, Practice and Methodological Underpinnings (M. Mehar and N. Prasad)
DOI: 10.1079/9781789249910.0009

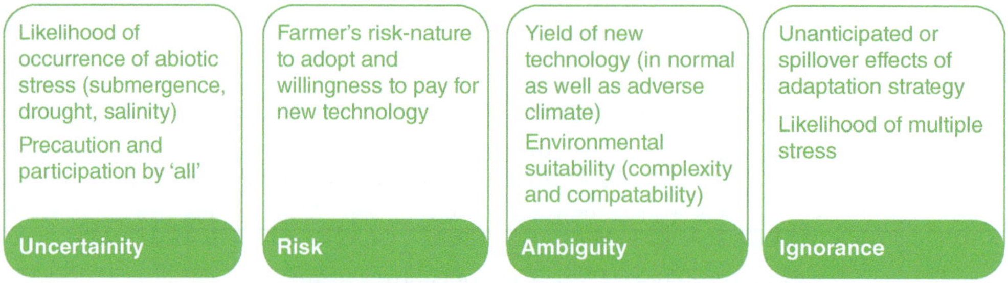

Fig. 9.1. Farmers' infinite pool of concerns. (From: the author)

First, there is **uncertainty**. The global and local concern that climate change is a continuous threat is not deniable by any individual. However, there is no surety about the likelihood of climate stress occurrence in the following crop season, as well as about the timing if it happens. Apart from this, farmers may not be fully aware of the entire pool of existing or latest CRS strategies, but they may have some knowledge about some. Their precaution and participation for climate mitigation are bound by their adaptive capacity and self-considerations; so, at the macro level, there is uncertainty whether everyone will adapt available strategies.

Risk relates to the farmers' decision and willingness to invest in CRS strategies due to their risk behaviour. The nature of such concern is determined subjectively by an individual and it varies according to their psychology and resource capacity.

Ambiguity concern is not influenced by probabilities but the problematic characterization of outcomes. For farmers who are usually bound by traditional beliefs and farming methods, technological change is a source of ambiguity concern as the yield is not predictable. One prime reason for farmers continuing to grow traditional varieties is that farmers have clarity about, and experience of, the adaptability and yield (in both normal and adverse situations). The other concern is 'environmental suitability' of technology, as often technology comes as a package and the agroecology condition and plot typology may not be conducive to the technical need.

Finally, there is the concern of **ignorance**, where 'neither probabilities nor outcomes can be fully characterized' (Stirling and Scoones, 2009). The adaptative strategies may have an unanticipated outcome or spill-over effect, as discussed in Chapter 8. Often farmers and practitioners ignore the aspect of the likelihood of more than one stress in one crop season and thus leave themselves unprepared to deal with the situation.

To understand how farmers decide these complex concerns needs a deep understanding of weather stress impacts on growth and productivity as well as timing and intensity of their occurrence. Each crop, as well as livestock and fish species, has a different sensitivity to different types of climate stress and associated vulnerability at specific stages of their (crop) development. Though the abiotic stress affects the village, the magnitude will vary for individuals and households due to other factors (discussed in Chapter 8) and hence there are different intra-household and inter-household views about adaptive strategies. Predicting climate stress occurrence at any point in crop development and deciding on action is determined by who does what activity and when they do it, and this is a complicated task. As the farming activities of men and women differ across context and different types of farming systems (or whether the farmers are well-off or poor) these may not be the same.

Although several studies and books have been produced to explore the first two concerns (uncertainty and risk) in Fig. 9.1, there is little evidence for the other two. Besides, the assessment of crop stage dimensions, discussed above, remains negligible.

This chapter elaborates on the conceptual framework presented in Mehar (2016). The framework encompasses all the activities and actors in the production cycle and their impact on food availability and government policies in the arena of climate event occurrence. Although the framework and discussion are limited to crops (and further limited to rice, as rice has a transplanting stage also), the issues and concerns raised are relevant to livestock and fish stages of rearing for sale or consumption. This chapter contributes to why gender mainstreaming is needed in CRS programmes conducive to each stage of crop/livestock/fish development.

9.2 Conceptual Framework

The conceptual framework presented in this section elaborates on the framework presented in Mehar (2016). The framework, based on livelihood diversification theory (Ellis, 1988) and the characteristics demand theory (Lancaster, 1966), aims to conceptualize how farmers make decisions about farming activities (such as selecting a variety(ies), deciding on the time of sowing, deciding land allocation to a particular variety) including climate-resilient strategies in a particular season.

According to this literature, livelihood diversification is the most widespread strategy. 'Livelihood diversification, which can be voluntary or involuntary, is defined as the process by which rural families construct a diverse portfolio of activities to survive and improve their standards of living' (Ellis, 1988). Growing a portfolio of varieties of crops is one of the most common strategies farmers adopt to address various concerns including risk (Wale *et al.*, 2011); e.g., as documented by FAO and ICRISAT (2004, p.20), 'by planting different varieties of cereals and legumes, farmers benefit from more productive but less hardy varieties, while at the same time hedging this risk with varieties that are less productive yet more tolerant to drought'. Moreover, farmers often either change the sowing time of a variety in their different plots or cultivate early- and late-maturing varieties of the same crop to spread out the amount of work required at harvest time.

The characteristics demand theory considers that 'goods are not desirable by themselves, but only in so far as they possess characteristics which provide utility for the consumer' (Lancaster, 1966). The characteristics model helps us predict how and why farmers' preferences will change with the availability of climate-resilient strategies due to the respective attributes of the adaptation strategy. The theory postulates that farmers or farm households maximize their utility by using several practices to benefit from their different attributes rather than selecting based on visible attributes. For example, different crop varieties are recommended on the grounds of yields, disease resistance, maturity time and test weight, but with varying weights or values of these features etc. However, other characteristics being equal, varieties that have had the highest average yield over years are primarily considered by the farmer. The choice of variety is a revealed preference[1] in responding to production constraints, satisfying consumption preferences, and fulfilling specific market requirements (Smale *et al.*, 2001).

Fig. 9.2 is a diagrammatical representation of a conceptual framework that sketches the links among farmers' characteristics and decisions regarding climate-resistant strategies. It portrays the mechanism with two approaches: (a) general framework discussing the crop cycle in the context of crop; and (b) characteristic model contextual characteristics influencing CRS decisions (here crop variety) of adoption and extent of variety adoption at the time of sowing.

9.2.1 Prior conditions and context

One premise of the livelihood literature is that farmers' activities and decisions are bound by policy, environmental and resource constraints. The impact of a climate shock, as well as mitigation strategies for climate stress at the macro level, depends on village-level factors (Rathore, 2005) such as village endowment, infrastructure and overall development and land characteristics of the village. Villages, where there are active private or public initiatives for disseminating information and training in advanced technology, are more likely to have improved farming methods and adopt new technologies.

The priority conditions and context (of policy settings, agroecology, demography, enabling infrastructure and market conditions, resource endowments and socio-economic conditions) influence farmers' resource allocation behaviour and decisions about livelihood strategies. It is a good idea to start with knowing the farmer's type of natural resource livelihood dependency that is important for household food security and income. Most often, the first question a researcher asks a household is: 'What is your primary and secondary source of income or livelihood?' (options: agriculture, livestock, forestry, fishery, or any combination of these). The answer does not exclusively exist in the household dynamics but on the implicit supporting environment and macro-economic conditions. For example, as mentioned in Chapter 6, in coastal areas rural people depend on fisheries for their livelihood. Additionally, government policies and programmes (e.g. subsidies, irrigation facilities) create an enabling environment for the livelihood source or specific crops, livestock or fisheries in particular. Markets (suppliers of inputs and buyers of harvest) also set enabling or barrier conditions for technology and practices. Market and infrastructure development are therefore important for successful envisioning of new farming practices.

Climatic stress perception is likely to find its roots in past events (Weber, 2006), experienced by individuals as well as the community. People's previous climate change experience, which may have affected them badly, strongly influences their farming decisions including which crop in the current season.

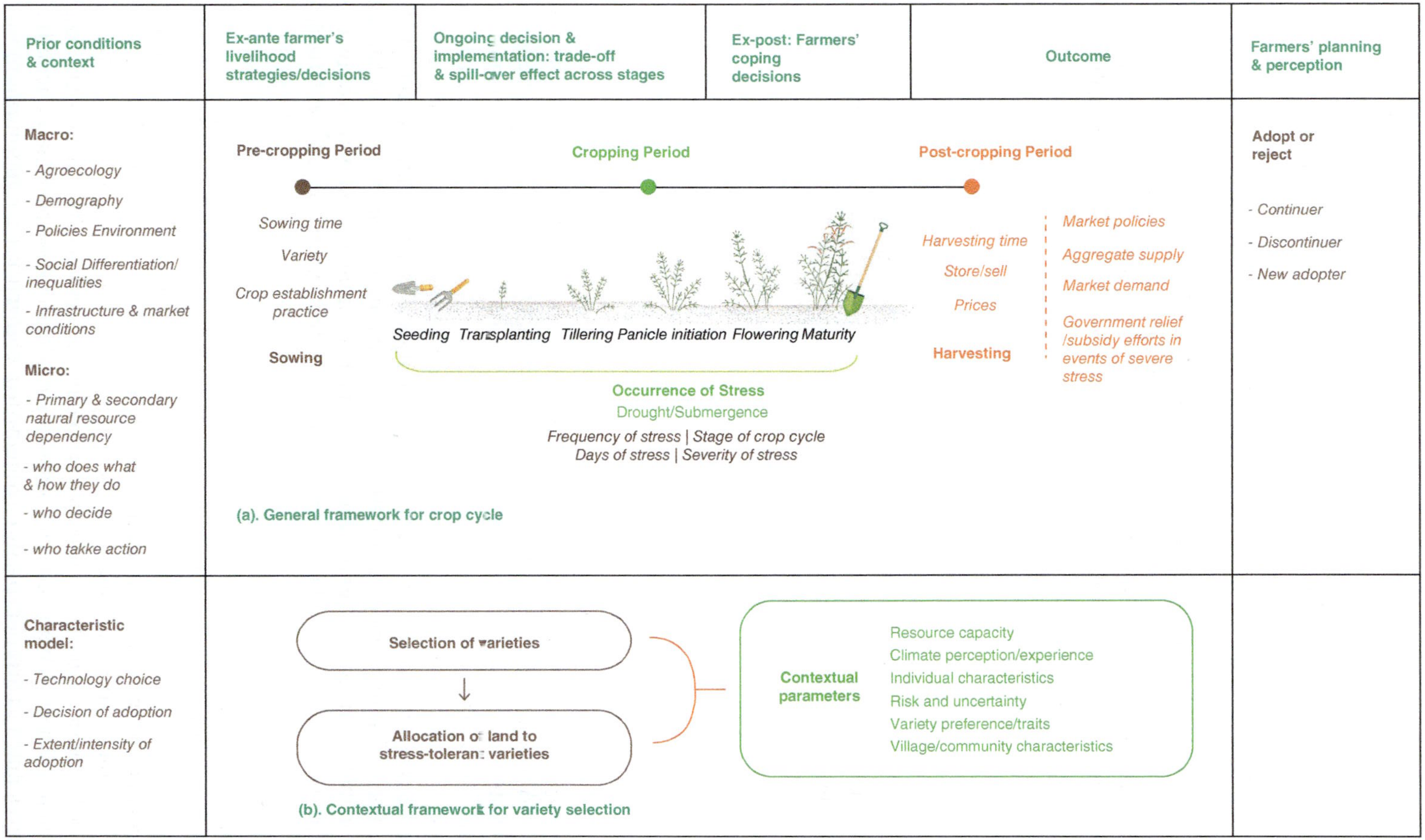

Fig. 9.2. Conceptual framework depicting: (a) variety outcome in general crop cycle; and (b) varietal selection decision at sowing time. (Original source: Mehar, 2016)
NB: transplanting is in case of rice crop only.

It is important to remember the micro-economic condition, influenced by the macro condition of social norms as well as intra-household gender relations and bargaining power. The combination of activities planned for the crop season by the farmer can be seen as a 'livelihood portfolio', as defined by Scoones (1998). In each setting or household, who does what and when, who decides and who acts, have a role in determining the livelihood portfolio. In the micro context, the next important thing to know is the type of crop, as well as a specific variety of crop grown by the farmer, as each crop, livestock or fish is sensitive to different weather conditions. To clarify this argument, we discuss rice and climate change linkage. Rice is selected to take advantage of learning about climate change in its additional activity of 'rice transplanting', which is not the case with other crops.

9.2.2 Rice and weather

Rice is a climate-sensitive crop and is highly dependent on irrigation with a high water requirement. So for land-preparation till transplanting, known as germination or early vegetative growth (20–25 days), low rainfall, late rainfall and drought could be a problem. Following China, the other top ten major rice-producing countries are developing where monsoons and glacial rivers are the main sources of irrigation. These developing countries predominantly use manual transplanting for rice. The other method is broadcasting and direct seeding. In manual transplanting, rice seedlings are grown in a small area (also known as a paddy field) for the first 15–40 days depending on the variety-nature and then transplanted with a uniform gap into a large area. This is a very laborious and time-consuming activity. In most rice-growing areas, women are often responsible for this activity. The main disadvantage of this method is the emission of methane gas from flooded paddy fields. Paddy rice contributes 10% of total agriculture greenhouse gases (GHGs), globally.

Although various studies measure climate change in terms of an increase in CO_2 and temperature, which have a positive impact on rice yield (see Aggarwal and Mall, 2002; Krishnan *et al.*, 2007), other studies have documented the negative impact of climatic uncertainties such as droughts and floods, or submergence, on rice yield. Chapter 5 discussed the rice farmer's perception, experience and severity of drought and submergence subdivided by cropping stages and number of days. Overall, climate shocks (flood and drought) will destroy the entire crop irrespective of the crop stage at which they occur. For climate variables (like low or late rainfall or submergence), the magnitude of the effect depends on at which stage they occur. During the flowering (15–20 days) stage, the effect will be more from stress from heatwaves and drought. At early grain filling/grain maturation/harvest (45–60 days), the most severe problem is frost followed by extreme events (drought and flood). So it is very crucial to decide the rice variety (normal, drought-resilient, submergence resilient, etc.), depending on the geographical and tropical context of the area.

For the most part, practitioners and policymakers fail to understand the linkage of climate change and crop development stages for sustainable agricultural development. This could partly be because the research, until recently, was mostly done by people from a different discipline who lack the opportunity to benefit from the other discipline. The other reason could be less focus on participatory approaches. Farmers and researchers may have a knowledge gap and may perceive things differently.

9.2.3 *Ex-ante* decision at the pre-crop period

Farmers' planning at the beginning of the crop is based on the concerns raised in Fig. 9.1. At time t_s, the farmer sows the variety; this may be a time horizon, as discussed earlier – farmers often plan different timings of sowing in different plots. The harvesting is done at time t_h, which again will be a time horizon based on sowing and climate conditions at that time. Between t_s and t_h, there are different stages of crop, i.e. tillering, panicle initiation, flowering and crop maturity. Farmers' planning has to be perceptive: (i) farmers' climate-stress perception; (ii) livelihood-diversification perception.

The climate problem is nebulous; in the words of Knight (1921) we are in a world of uncertainty (unknown probability) rather than risk (known probability). The occurrence of climate stress (say drought and submergence) between t_s and t_h is uncertain. The frequency (number of times stress occurs, number of days it continues), its severity (effect on overall crop yield, the effect on variety-wise yield), and during which crop cycle they occur is difficult to predict at t_s.

Farmers' stress perception may be influenced by different aspects of climate event:

- Timing, extent and type of climate stress: like rice, other crops, as well as animals, have different sensitivities to weather stresses at different stages of their development. The degree to which a stress's type, duration, periodicity and at what point in the crop development stage it occurs and is identifiable makes a huge difference to the plan of adjustment or adaptive strategies and the end outcome.
- Frequency: stress in combination (two or more) and appearance at different stages of crop development has less likelihood of identifiable predictability and controllability yet more impact than an individual stress.

As explored in previous chapters, male and female farmers have different knowledge of crops and activities. Depending on their knowledge, who does what and when in the crop development stage may have a huge influence on the accuracy of the predictability of climate events. The predictability of a climate event, especially in accordance with crop sensitivity and stage of development, helps farmers decide the type and timing of sowing and using *ex-ante* CRS strategies.

Different livelihood portfolios have different pathways, timescales and specializations, with a concentration on one activity, or a limited range of activities available to the farmer. How individual farmers weigh their perception about the occurrence of climate change as well as the outcome of CRS is inevitably related to inequities in adaptive capacities and pre-existing social norms. Due to land constraint, agricultural extensification (i.e. adding more land for cultivation and more production) is less viable than agricultural intensification (gaining more production per unit area through either increasing technology (capital-led) or increasing labour input, or both) and coping strategies (such as migration, looking for a non-agricultural job, reducing food basket size). Depending on the strategy or combination of strategies, farmers experience direct as well as spill-over effects as discussed in Chapter 6.

Changes in rainfall patterns and temperatures have a direct effect on plants and increase susceptibility to pests and diseases, which, in turn, affect the mean and variance of total crop yield. Farmers facing productivity shocks from climate change possess a menu of potential strategies such as: changing planting timing; crop rotation; changing to varieties resistant to climatic stresses; improved soil- and water-management practices including conservation agriculture; or shifting to non-farm livelihood sources like borrowing, eating less, migration or employment in non-agricultural activities (Mehar, 2014; Mehar *et al.*, 2016). However, which strategies farmers pursue not only depends upon the actual effects of climate change but also on how they perceive and cognitively process their experiences and update their perceptions of climate risk. In this context, it is important to capture information regarding both the experience of farmers and their knowledge/perception about climate change. A farmer's attitude towards the adoption of climate-friendly technology will depend on their understanding and perception of, and threat from, climate change. The farmers who have experienced climate stress over the past few years and have faced significant losses (crop yield), and perceive that the severity of these events will increase in future, are more likely to adapt to mitigate risk.

Since it is unlikely that one strategy will address all the attributes that the farmer looks for, farmers pursue multiple strategies together or in sequence depending on the requirement of the strategy as well as the level of risk associated with the alternative options. The selection decision, along with the extent of adoption, is explained with 'characteristic approach' theory. For a decision about varieties, the characteristics model postulates that farmers or farm households maximize their utility by using several varieties to benefit from their different attributes rather than select on the basis of attributes such as the output or yield of a single variety (Smale *et al.*, 1998). Varieties are recommended on grounds of yield, disease resistance, maturity time, test weight, etc. However, other characteristics being equal, varieties that have had the highest average yield over the years are usually recommended. The choice of variety is a revealed preference in responding to production constraints, satisfying consumption preferences and fulfilling specific market requirements (Smale *et al.*, 2001). The probability that farmers choose to cultivate a certain crop variety depends on the key attributes of the farmers associated with that variety. Overall, the extent to which farmers adopt improved varieties and replace their landraces depends on the extent to which the new varieties better fit their household livelihood strategies (Maxted *et al.*, 2002). Farmers' reasons for less and partial adoption of improved varieties are primarily based on their trust in the local varieties in terms of yield, better adaptation and maintaining soil quality.

> Like this 'seed variety' example, farmers have similar concerns for all other CRS strategies.

It is important to mention that farmers' contextual characteristics (age, gender, education, religion, caste, family size or dependency ratio, risk-averse nature, impatience level) influence their attitude to adopt new technology. When a farmer chooses to adopt a new variety and replace an older variety, it reflects the farmer's judgement that the new variety offers some net benefit or advantage (Evenson and Gollin, 2003). The psychology of an individual about risk is also affected by gender, equity position and family situation. A farmer with a family must provide for them (if it suits his value system) and is often forced to be conservative (Walker, 1959).

The gender of a farmer is another crucial determinant of seed selection. While men generally focus on criteria related to agronomic characteristics and market value, women apply additional criteria related to food consumption such as palatability, taste and cooking quality. Furthermore, men and women share criteria related to the quality of seed, such as seed size and freedom from pests (FAO, 2007, quoted in Loconto and Shapiro, 2008).

9.2.4 Post-cropping period (*ex-post*-event occurrences)

There is increasing evidence of heatwaves occurring during crop-maturity stage, which affects the total crop production and farmers' planning (early harvesting and, if perceived at crop-planning stage, use of short-duration varieties or early sowing so that early harvesting can be done). In the case of huge production loss as a result of climate shock, a farmer tends to adopt generic coping strategies as mentioned in Chapter 6.

Post-harvest activities of cleaning, drying, winnowing, sorting, packing, processing, storing and marketing are, apart from marketing and storing, predominantly viewed as female responsibilities. Marketing, in the case of crops, is considered a male activity, and in the case of fish (especially marine fish), a female activity.

There is limited knowledge about post-harvest losses, their contribution to GHG and gender issues in post-harvest management. Grewer *et al.* (2018) have measured the effect of post-harvest losses on GHG emission intensity for several crops and livestock. The gender roles and relations pre-existent due to social norms affect the activities in grain storage management. Even for crops considered to be 'male' crops, the farming decisions, like which variety to grow, might be strongly yet indirectly influenced by women, as they have important roles in post-harvest management, food security (storing grain for the household), consumption, and cooking for household, thus better knowledge.

9.2.5 Outcomes

The activities and actors in the agricultural system lead to outcomes such as total production of the crop, market supply and price, food security and vulnerability (Fig. 9.2 – extreme right). Depending on the occurrence and severity of stress, the harvest scenario for individual and community level will be affected. The aggregate harvest is likely to be low as the severity of stress increases and, as a result, prices may likely rise. So if the stress is moderate and farmers have moderate yields, this, coupled with a resultant higher market price, may lead to normal profits without a high reduction in expected profit. Crop losses due to climate change have implications for the policy as well, since in cases of severe drought/heavy floods, relief is provided by government and NGOs, which influences the actual impact of stress on farmers' livelihoods. Moreover, climate stress will be more severe for some sections of society (e.g. landless women and children). This segment is doubly affected: (i) they have fewer resources to cope with risks immediately; and (ii) since they usually form a significantly large segment of the economy, their forced dependence on the market to meet their food-requirement gap in poor crop years is likely to push food prices higher, which is likely to hurt them the most. In addition to crop losses, another prolonged effect might be related to health and sanitation issues, especially in cases where water pollution has increased.

9.2.6 Future planning and perception

The end-results of rural welfare (incomes, productivity, poverty, vulnerability, etc.) and natural resources outcomes are sustainable or unsustainable. As climate conditions continue to change, individual agricultural activities and entire farm systems will also need to change. Climate change is a process without end, so adaptation will be continuous.

For a given CRS, farmers can be categorized into three categories:

1. Continuer: farmers who participated in the previous crop season and continue to participate or adopt the strategy in the current season. Rogers (2003), using the concept of individual innovativeness theory, categorizes these continuers based on who adopts an innovation (adaptive strategy) and when (Box 9.1).
2. Discontinuer: farmers who participated in the previous crop season and reject the strategy for the current season.
3. New: farmers who are planning to use the strategy for the first time in the current season.

The reasons for continuing or discontinuing, examined in a few studies, can provide important feedback loops regarding whether there is a need for

Box 9.1. Adopter or continuer categories.

Innovators – risk-takers and pioneers who lead the way.
Early adopters – these follow innovators and communicate about the innovation to others. They include the progressive farmers, large landholders, etc.
Early majority – those who adopt the innovation after getting informed and getting informal assurance about the sustainability of the innovation by their formal and informal social contacts including the adopters from the first two categories.
Late majority – highly risk-averse and wait to make sure that the innovation is in their best interests.
Laggards – most have poor adaptive capacity and are usually from the lower strata of the society.

revising adaptive strategy of the dissemination process, and if yes, how. The reason influencing the decision is due to the perceived attributes (Rogers, 1995):

- Compatibility: the degree to which the strategy is perceived as consistent with environmental suitability, as well as the needs of potential adopters, influences the chances of (dis/)continuation.
- Complexity: whether the strategy is easy to understand and use – most farmers are uneducated and thus may have difficulty in understanding the technical part.
- Relative advantage and observability: the adopter will continue it if they perceive or experience the benefits (yield gain or low-input need) to be more than their traditional method.
- Trialability: the degree to which an innovation may be experimented with on a limited basis. CRS are tried and experimented on the field before communicating them to others. The adoption success or constraints are to some extent visible during their trial at different farming stages.

9.3 Summary

This chapter shares basic facts, concepts and principles related to livelihood diversification pre- and post-climate events. The discussion here is not intended to serve as a comprehensive desk reference but brings attention to the importance of assessing farmers' climate-resilient strategies including the perception of climate as it relates to each stage and to different crops. The chapter also discusses the importance of gendered roles and relations in each crop stage and echoes the need for gender mainstreaming at each stage.

Note

[1] Revealed preference theory, initiated by Samuelson (1938, 1948), is a method of analysing choices and preferences made by individuals. For example, if the consumer has two available choices (*x*, *y*) and the consumer prefers *x* while *y* is available (maybe because *x* has better quality or price features than *y*), then s/he reveals her/his preferences for *x* over *y*. Equivalently, we say *x* is a revealed preference over *y*.

References

Aggarwal, P.K. and Mall, R.K. (2002) Climate change and rice yields in diverse agro-environments of India. II. Effect of uncertainties in scenarios and crop models on impact assessment. *Climate Change* 52, 331–343.

Ellis, F. (1988) *Peasant Economics: Farm Households and Agrarian Development*, 2nd edn. Cambridge University Press, Cambridge, UK.

Evenson, R.E. and Gollin, D. (2003) Chapter 3 – Objectives and methodology: crop genetic improvement studies. In: Evenson, R.E. and Gollin, D. (eds) *Crop Variety Improvement*

and Its Effect on Productivity: The Impact of International Research. CAB International, Wallingford, UK, pp. 7–38.
FAO (2007) "Seeds are my wealth. Men and women farmers and their seeds in Tanzania." CD-ROM.
FAO and ICRISAT (2004) The impact of HIV/AIDS on farmer's knowledge of seed: Case study of Chókwe District, Gaza Province, Mozambique. Research report. Available at: https://reliefweb.int/report/mozambique/impact-hivaids-farmers-knowledge-seed-case-study-ch%C3%B3kwe-district-gaza-province (accessed 21 March 2022).
Grewer, U., Nash, J., Gurwick, N., Bockel, L., Galford, G. *et al.* (2018) Analyzing the greenhouse gas impact potential of smallholder development actions across a global food security program. *Environmental Research Letters* 13(4), 044003.
Knight, F.H. (1921) *Risk, Uncertainty, and Profit*. Houghton Mifflin Company, Boston, Massachusetts.
Krishnan, P., Swain, D.K., Bhaskar, B.C., Nayak, S.K and Dash, R.N. (2007) Impact of elevated CO_2 and temperature on rice yield and methods of adaptation as evaluated by crop simulation studies. *Agriculture, Ecosystem and Environment* 122(2), 233–242.
Lancaster, K. (1966) A new approach to consumer theory. *The Journal of Political Economy* 74, 132–157.
Loconto, A. and Shapiro, B. (2008) *Diversity of experiences: understanding change in crop and seed diversity: a review of selected LinKS studies*. Gender, Equity and Rural Employment Division. Available at: https://hal.inrae.fr/hal-02821201/document (accessed 2 May 2020).
Maxted, N., Guarino, L., Myer, L. and Chiwona, E.A. (2002) Towards a methodology for on-farm conservation of plant genetic resources. *Genetic Resources and Crop Evolution* 49(1), 31–46.
Mehar, M. (2014) Gender specific impacts of climate variation in agriculture. Unpublished MPhil thesis, Department of Economics, School of Social Science, Indira Gandhi National Open University, India.
Mehar, M. (2016) Role of gender differentiated risk and time preferences in rice varietal selection in eastern India. PhD thesis, Department of Economics, School of Social Science, Indira Gandhi National Open University, India. Available at: https://shodhganga.inflibnet.ac.in/handle/10603/212284 (accessed 21 March 2022).
Mehar, M., Mittal, S. and Prasad, N. (2016) Farmers' coping strategies for climate shock: Is it differentiated by gender? *Journal of Rural Studies* 44, 123–131.
Rathore, M.S. (2005) *State level analysis of drought policies and impacts in Rajasthan*, India. IWMI, Colombo, Sri Lanka. Working Paper 93. Drought Series: Paper 6.
Rogers, E.M. (1995) *Diffusion of Innovations,* 4th edn. Free Press, New York.
Rogers, E.M. (2003) *Diffusion of Innovations*, 5th edn. Free Press, New York.
Samuelson, P.A. (1938) A note on the pure theory of consumer's behavior. *Economica* 5, 61–71.
Samuelson, P.A. (1948) Consumption theory in terms of revealed preference. *Economica* 15, 243–253.
Scoones, I. (1998) *Sustainable Rural Livelihoods: A Framework for Analysis*. Institute of Development Studies, London.
Smale, M., Singh, R.E., Sayre, K., Pingali, E., Rajaram, S. *et al.* (1998) Estimating the economic impact of breeding nonspecific resistance to leaf rust in modem bread wheat. *Plant Disease* 82, 1055–1061.
Smale, M., Bellon, M. and Aguirre, A. (2001) Maize diversity, variety attributes, and farmers' choices in Southeastern Guanajuato, Mexico. *Economic Development and Cultural Change* 50(1), 201–225.
Stirling, A.C. and Scoones, I. (2009) From risk assessment to knowledge mapping: science, precaution, and participation in disease ecology. *Ecology and Society* 14(2), 14.

Wale, E., Drucker, A.G. and Zander, K.K. (2011) *The Economics of Managing Crop Diversity On-farm: Case-Studies from the Genetic Resources Policy Initiative*. Earthscan Publications, London.

Walker, O.L. (1959) Game theory applications in agricultural decisions. Retrospective theses and dissertations, Iowa State University.

Weber, E.U. (2006) Experience-based and description-based perceptions of long-term risk: why global warming does not scare us (yet). *Climatic Change* 77 (1–2), 103–120.

10 Gender, Climate Change and Sustainable Farming Systems

Abstract

For sustainable food and production systems, production and availability of healthy and nutritious food is essemtial and can be obtained by optimizing resources across agricultural sub-sectors (crops, forestry, livestock, fisheries, aquaculture). The need for exploring the trade-offs and opportunities in integrated agricultural sectors (crop production, forestry, fisheries, aquaculture or livestock) is twofold: first, the sectors compete for resources and their synergies offer potential for reduced emission of carbon sinks; and second, farming households are often involved in two of these sub-sectors for their livelihood. As to which is the primary and secondary sector as a source of livelihood, men's and women's participation, priorities and preferences vary.

10.1 Introduction

For a sustainable food system, our vision is a sustainable and resilient production system producing healthy and nutritious food which is accessible and available in equitable manner to everyone in the society. This is keeping in mind two major points: (a) the United Nations Sustainable Development Goal 2 target of achieving 'zero hunger by 2030'; and (b) the world is facing existential challenges such as rising population, changes in dietary preferences and climate change, threatening livelihoods and production systems. To achieve this vision, we need an approach which optimizes synergies as well as the allocation and management of resources among the common elements of the food systems (i.e. pre-production, production, supply chain, processing, distributing, marketing and consumption of food) and provides equitable opportunities and resources to crop farmers, pastoralists and livestock keepers, fisher farmers, foresters and other rural workers or entrepreneurs (who provide services connected to agriculture). With equitable opportunities, both rural men and women will be able to contribute to agricultural production and climate mitigation in their own and collective ways.

 Climate Change, Adaptation and Gender: Policy, Practice and Methodological Underpinnings (M. Mehar and N. Prasad)
DOI: 10.1079/9781789249910.0010

Past emphasis on mono-cropping or individual sub-sectors (crop production, forestry, fisheries, aquaculture or livestock) in agriculture has resulted in depletion of natural resources more than what it could have been if the sectors had worked together. This is a crucial reason for stagnation in agriculture growth, as the sub-sectors compete and conflict over time and space for natural resources as well as government support and policies. It is important to recognize and act on the risks, challenges and synergies across the sub-sectors as well as across the five areas identified above.

In the previous chapters, the discussion and examples are provided from all sub-sectors of agriculture. However, as highlighted in different parts of the book, there is a dearth of literature for sectors other than crop farming, so the argument to some extent remains crop-dominant. Further, the literature remains dominant on exploring individual sub-sectors, crowding out the possibilities of trade-offs and conflicts in sub-optimal allocation and management of resources. In this chapter, we discuss the positive and negative synergies for mitigating climate change across agricultural sub-sectors, highlighting the participation and contribution of men and women.

The uptake and scaling of climate-resilient practices across the sector will ensure long-term gains in productivity and food availability. The practices with entrepreneurial approaches could be viewed as an efficient way to assure farmers and thus could be viewed as a precondition for ensuring sustainable agriculture, rural development and food systems. As stated by Opolot *et al.* (2018), 'once the farmers are able to improve their household income and food security, the uptake of improved technologies will be hastened'. Moreover, climate change affects the quality of food production and also results in several health issues of the affected population, thus the health and nutrition dimension for a sustainable food system is a prerequisite.

As a concept, sustainable agriculture better encapsulates the three pillars: cross-sector integration and synergies, climate-smart practices with an entrepreneurial mindset, and effect of climate change and climate-resilient practices on the health and nutrition of rural people.

10.2 Cross-sectoral Integration and Synergies

As illustrated in Fig. 1.1 (Chapter 1), climate change is affected by agricultural activities in many ways. Globally, around 25% of greenhouse gas (GHG) emissions come from agriculture, forestry and land use alone. In Chapter 2, Fig. 2.2 illustrated the direct and indirect effects of agricultural activities on climate change as well as the impact of climate change on agricultural activities.

The discussion was for each of the sub-sectors of agriculture separately; however, this 'excessive separation has often created a "silo syndrome" in which crop production, livestock, forestry, capture fisheries and aquaculture compete with each other for space, political support and natural resources, often resulting in conflicts and in sub-optimal allocation and management of resources' (FAO, 2014, p. 15). As the supply of land is inelastic, there is often competition for land among crop, livestock and aquaculture sectors which, in turn, results in deforestation. The other common component of competition is freshwater use in all sectors.

There are also potential and positive synergies across sectors. For example, the crop sector provides forage and fodder to livestock, and feed to aquaculture. Forests provide shelter and food to aquaculture, fisheries and livestock. Livestock provides manure for crop farming and thus there is less need of mineral fertilizer. These positive synergies suggest possibilities of integrated production systems where they act together and optimize use of available and limited resources. In integrated systems the adaptive capacity of farmers is influenced by the nature and extent of trade-offs between the components of the farming system and their degree of integration (Dixon *et al.*, 2014 quoted in FAO, 2017). A few of the integrated systems are:

1. **Integrated crop–livestock system.** Crops provide fodder and plant residues to livestock; thus, in this system, emissions from the disposal of crop residues and by-products are reduced. In turn, livestock produces manure, which is an organic fertilizer and beneficial for crop productivity. Manure can also be used to produce bio-energy which can be used for home consumption as well as a replacement for fossil fuels. This use of manure reduces GHG emissions. Livestock also supports farming activities like ploughing, transporting material, etc. A specific example of this system is **agropastoral**, which is found in pastoral and agropastoral areas and is associated with dry land or rainfed crop production.
2. **Integrated forestry.** All these sectors (crop, livestock, fisheries, aquaculture) are 'supported by forests, which provide ecosystem services, especially soil formation, water purification, biodiversity conservation and climate regulation' (FAO, 2014). The integrated system with different sectors is of different types. These include: agroforestry (trees on farm land or plots), aquaforestry (trees lining fish ponds) and apiculture (the practice of keeping bees); silvopastoral systems (integration of trees and shrubs in pasture with animals); and agrosilvicultural systems (integration of woody perennial trees with crops). The trees integrated in these systems are used as shade trees, shelterbelts and windbreaks, etc.
3. **Integrated crop–fish system.** A common example of an integrated crop–livestock system is rice–fish farming. A special case of this system is **aquaponics,** in which aquaculture (the practice of fish farming) and hydroponics (the cultivation of plants in water without soil) are combined.

Likewise, there is potential for other integration systems like integrated fish–livestock systems (e.g. duck–fish farming and poultry houses made above ponds). Fig. 10.1 provides examples of climate-resilient strategies in solo and integrated systems. 'Integrated systems can play a critical role in mitigating greenhouse gases from agriculture, as their emission intensities are typically lower than the sum of those from specialized systems.'[1].

Soil or land, and water (including natural resources, monsoon and ground water) are shared resources across these agricultural sectors. Because of this, the projects and policies implemented related to soil or water conservation may have spill-over effects, positive or negative, or may be both for other sectors.

Similar to crops, there are gender diversities in roles and activities performed in different stages of livestock, capture fisheries, aquaculture and forestry. Among all these sub-sectors, women's role is found to be more dominant and more visible in livestock as compared to others. According to FAO, 70% of those employed in the livestock sector are women.[2]

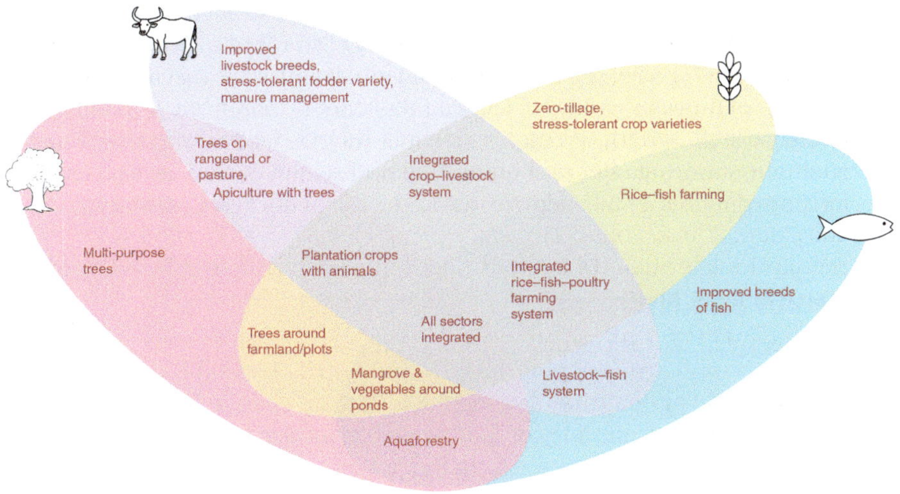

Fig. 10.1. Examples of climate-smart agriculture within crop, livestock, fishery, agroforestry and integrated systems. (From: the author)

In capture fisheries and aquaculture, women account for more than 19% of people directly engaged in 2014 (FAO, 2016). In a few of the activities, women's role remains dominant, such as post-harvest processing of capture species (such as drying and salting of fish) or fishing methods (such as gill nets, dragnets, harpoons, etc.) (Jiddawi and Ohman, 2002). Pond tenure in the aquaculture sector is also highly gendered. Women tend to be managers of backyard aquaculture ponds, which is in keeping with the social norm of restrictions on their mobility (FAO, 2017), whereas in forestry, 83% out of the 850 million rural poor reliant on forests for collecting wood for fuel, medical plants and other resources for food for the family consumption are women (FAO, 2018).

Rural women and men have distinct tasks, control and perception of the way they use agricultural resources. In general, women have more control over the income produced by the sale of livestock products, vegetables from home gardens and residue crops from fields (such as fodder shrubs, herbaceous legumes and improved grasses). In agroforestry, women take advantage of indigenous fruits and vegetables, and the gathering of fodder and mulch. They also have indigenous knowledge of local crops, trees, herbs and wild edible plants in forests. Gender-responsive climate-smart activities can include tree species that provide wood for fuel, fodder, shade and fruit to meet the demand of women, and timber to meet the demand of men. Other possible interventions aimed at reducing the gender gap in agroforestry are enhanced feed management for livestock and aquaculture species, community-based small-scale fisheries management, and mangrove forests management, supporting diversification.

In integrated systems, some sectors are found to be in favour of one gender. For example, in crop–livestock production systems, women are found to be more involved in livestock production. They clean cowsheds, milk the cows, look after calves and sick animals, cut the grass, supervise feeding and grazing of cows, and make dung cakes, butter and cheese. Women sell milk and other animal-based products locally and have full control over the income earned.

Similarly, in crops and aquaculture, men and women have different roles. For example, in a focus group discussion in Odisha (India) and Bangladesh, men were found to be working in rice farming and females taking care of ponds. 'Male gives more times to paddy, so female takes care of ponds' (a respondent, quoted in Mehar *et al.*, 2020). Women's activities include feed management and water-quality monitoring, whereas harvesting and selling of fish, as well as buying of fry/fingerlings, is a male activity due to the social norms of mobility.

10.3 Health and Nutrition

Production and availability of healthy and nutritious food is an important pillar of sustainable agricultural development. Climate change has an adverse impact on the quality and quantity of food available, as discussed in Chapter 2. Further climate changes will have a strong impact on the health and nutrition of rural men and women. In a climate shock situation, the immediate impacts are mortality, morbidity and disability (Fig. 10.2). The short-term consequences of climate change are impacts on crop failure or low production, thus low availability of food – thus, the triple burden of nutrition/hunger, undernutrition and malnutrition. The impacts are different for men, women, girls, boys or the elderly. A common example, seen in Chapter 8, is reducing food intake by female members of the family. However, the impact is not limited to food intake issues but has other dimensions which can have a direct impact on health, such as sanitation, quality of water, and diseases (food-borne, vector-borne, water-borne and others) (Tirado *et al.*, 2010; Phalkey *et al.*, 2015; Fanzo *et al.*, 2017; Myers *et al.*, 2017).

Climate change also affects food supply and prices, in turn impacting the overall income and purchasing power of farmers and thus limiting access to nutritious food such as vegetables, fruit and animal-source foods. Some of the coping strategies, like selling assets (i.e. livestock, jewellery, etc.), may have an immediate impact on food availability but have long-term negative effects on household equity. Though not much assessed, mental health issues, especially in cases of climate shock or as a consequence of forced migration or domestic violence, etc., are also a significant factor to be considered.

Gender-differentiated impacts exist due to the different responsibilities of men and women. Female members of families are viewed as natural custodians of

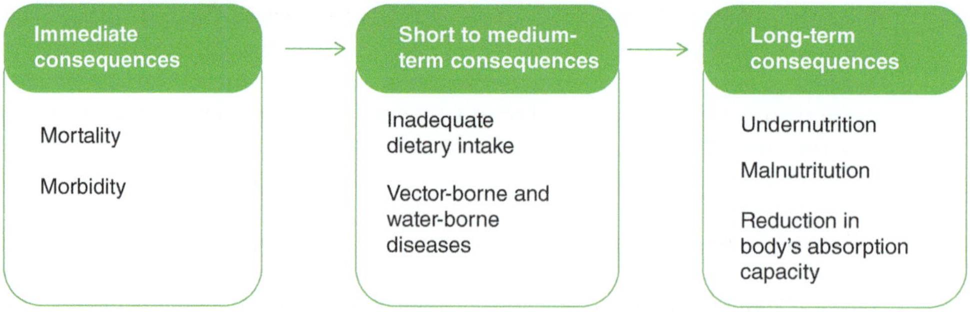

Fig. 10.2. Impact of climate change on health and nutrition. (From: the author)

household food security, food consumption and nutritional security. Their prime activities are cooking and distributing food, and other activities such as seed storage, taking care of livestock and thus dairy products. In a few cases, rural women also grow vegetables in small spaces in their homes, which adds to the nutritional intake of the family. Forestry women collect wild plants for protein and medical purposes. It is important to note that not only production – and thereby consumption – determines good health; how the food is prepared and consumed also have an effect on health and can mitigate the impacts of climate change (Fanzo *et al.*, 2017, p. 94). Furthermore, the effect of climate change in a pandemic puts double pressures on women to adopt CRS and secure food and nutrition.

Evidence suggests that female inclusion in CRS programmes is important and has a positive impact on improving household dietary diversity. For example, Jones *et al.* (2014) found that female-headed households in Malawi had increased dietary diversity (measured by household dietary diversity score and household food consumption score in a combination of livestock and crops). FAO (2011) estimates the yield gap in men's and women's productivity average around 20–30% and found that bringing the yield of land farmed by women up to the levels achieved by men would have productivity gains of 2.5–4%; and, globally, a 12–17% decline in undernourished people. Nelson *et al.* (2012) argued that if the effects of climate change were also considered, it is possible that these productivity gains would be even greater. The report also recommends 'designing policies and programs that target food security programs generally and climate-change-specific activities to women' to achieve both efficiency and welfare benefits. An empirical study in Ethiopia by Teklewold *et al.* (2019) found that the gender-disaggregation results suggesting nutritional outcome differentials between male- and female-headed households are due to differences in household characteristics, including household resources and differences in returns to resources. The authors also discuss the differences for climate-smart agricultural practices adopted in isolation or in combination.

Fanzo *et al.* (2017) identified focal areas for interventions to reduce nutritional risks of climate change along the food value chain. In a few cases, CSA demonstration and implementation are used to educate men and women about the nutritional benefits of practice, along with reducing climate risks. For example, Beuchelt and Badstue (2013) used a couple-based approach to identify what people like or dislike in relation to gender roles, their effects and possible solutions, and later integrated these into workshops with cookery demonstrations and nutrition education. An impact assessment of the Gender Oriented Participatory Extension Approach (GPEA) showed positive changes such as families' access to protein-rich food with the introduction of legumes, improving relationships between men and women and increasing women's empowerment (in terms of making their own decisions, active participation in workshops, keeping control over their income, etc.).

10.4 Entrepreneurship

The term 'sustainability' means continuous and certain growth over time which also needs timely investment in the sector. In agricultural economies, the majority of farmers are found to be small and medium-sized landholders. Two factors that remained constant over time in these economies were: (i) the majority of farmers were in debt; and (ii) lack of continued and significant investment in agricultural policies, infrastructure and markets. The situation is worse for farmers in climate-affected areas.

Opolot *et al.* (2018) suggest that 'for agricultural transformation to occur in developing countries necessitates that farmers are enabled to assume the mantle of their own development'. The authors further explore that promoting rural entrepreneurship is critical in contributing to the development of engagement with impacts and agricultural sustainability. Climate-resilient practices thus should be seen as viable ventures ensuring profits in addition to providing environmental services or as a pathway to achieve green growth.

However, entrepreneurship has been neglected in the rural climate adaptation literature to date (Akinbami *et al.*, 2019), let alone the discussion regarding the gender–entrepreneurship–climate nexus.

There are potential opportunities associated with men's and women's entrepreneurship for adapting practices with solutions to reducing climate risk.

Some of the climate-smart technologies (such as zero tillage (ZT) and laser land levelling (LLL)) are expensive, so most of the farmers use rental services. The service providers are relatively large entrepreneurial farmers who own one or more such technologies. The technologies also need technical knowledge and skills. The service providers are male-dominant entrepreneurs.

Women have key roles in community natural resources management, innovation, farming and care-giving. Their entrepreneurial activities can also help to eliminate some societal problems, such as new businesses that support CSA-related activities such as starting agro-input dealerships that sell stress-tolerant seeds or other inputs. In India, a non-pofit organization, Swayam Shikshan Prayog, created 5500 self-help groups to support women to become farmers, entrepreneurs and leaders under the one-acre model in Punqe, India (Yadav and Lal, 2018). These women practise climate-smart practices such as drip irrigation, tree plantation, use of biopesticides, etc. (ibid.). In Bangladesh, the Grameen Shakti (GS) microloans initiative combines renewable energy, jobs and skills training, which has helped to install more than 100,000 solar home systems in rural communities, creating employment opportunities while also empowering women and local youth (IPCC, 2007). Ironically, most of these micro-finance programmes designed for women are used and controlled by their male counterparts. Alam *et al.* (2015) found a similar example, where an NGO provided Luici Lights, solar lanterns that replaced kerosene lamps. This also saved women time in chopping down trees and gathering firewood, and the authors found women shifting to entrepreneurial ventures such as beekeeping or bead-making. In another case, Gram Mooligai Company Limited (GMCL), in four southern states in India, formed and managed by *untouchables* (backward caste), was the first female community enterprise in India active in the herbal sector and has empowered several females and supported them in other entrepreneurial activities (Torri and Martinez, 2014). For example, 19 out of 42 of the women interviewed were able to acquire goats and poultry (ibid.).

10.5 Summary

To ensure we are transiting towards the vision of a sustainable food system in existing climate-adverse situations, appropriate action should be identified and implemented within and across the three pillars of sustainable food systems identified above. A few more cautions and suggestions added below could be beneficial:

1. Simultaneous efforts are needed across all sectors to maximize benefits across objectives and minimize trade-offs across sectors and within.
2. Multiple interactions of partners (farmers, government, private players, etc.) and feedback loops throughout the system are important. Contextualization of issues will be feasible with a 'transdisciplinary' approach to policymaking. The latter needs decentralization of capacity and decision making across partners as well as a vertical-hierarchy government system (or multi-level transdisciplinary governance).
3. Initiatives are needed to break the blockages in some challenges.
4. Social innovations in food systems are needed. Diversification of crops, reducing waste and localization of public procurement systems will be beneficial to achieve a sustainable food system.
5. Linking local, national and international research would help in scaling up the technologies. For this, science-based evidence is needed to complement traditional knowledge of farmers in a participatory manner.
6. To date, there is lack of evidence to provide solutions for many of the issues in agriculture. Efforts are needed to get multiple forms of knowledge and interpretation of data/information to make informed and evidence-based decisions.

Notes

[1] http://www.fao.org/climate-smart-agriculture-sourcebook/production-resources/module-b5-integrated-production-systems/b5-overview/en/ (accessed 4 June 2020)

[2] http://www.fao.org/rural-employment/agricultural-sub-sectors/livestock/en/ (accessed 22 March 2022)

References

Akinbami, C.A.O., Olawoye, J.E., Adesina, F.A. and Nelson, V. (2019) Exploring potential climate-related entrepreneurship opportunities and challenges for rural Nigerian women. *Journal of Global Entrepreneurship Research* 9, 19.

Alam, M., Bhatia, R. and Mawby, B. (2015) *Women and Climate Change: Impact and Agency in Human Rights, Security, and Economic Development.* Georgetown Institute for Women, Peace and Security (GIWPS). Available at: https://www.preventionweb.net/publications/view/46922 (accessed 2 February 2020).

Beuchelt, T.D. and Badstue, L. (2013) Gender, nutrition- and climate-smart food production: opportunities and trade-offs. *Food Security* 5, 709–721.

Dixon, J.L., Stringer, L. and Challinor, A.J. (2014) Farming system evolution and adaptive capacity: insights for adaptation support. *Resources* 3(1), 182–214.

Fanzo, J., McLaren, R., Davis, C. and Choufani, J. (2017) *Climate Change and Variability: What Are the Risks for Nutrition, Diets, and Food Systems?* IFPRI Discussion Paper. IFPRI, Washington, DC.

FAO (2011) *The State of Food and Agriculture 2010–2011. Women in Agriculture: Closing the Gender Gap for Development*. Food and Agriculture Organization of the United Nations, Rome.

FAO (2014) *Building a Common Vision for Sustainable Food and Agriculture: Principles and Approaches*. Food and Agriculture Organization of the United Nations, Rome. Available at: http://www.fao.org/3/a-i3940e.pdf (accessed 23 March 2022).

FAO (2016) *Mid-term Evaluation of the Conservation-Agriculture Scaling-Up Project*. Office of Evaluation. Food and Agriculture Organization of the United Nations, Rome. Available at: http://www.fao.org/3/a-bq888e.pdf (accessed 23 March 2022).

FAO (2017) *Climate Smart Agriculture Sourcebook*. Food and Agriculture Organization of the United Nations (FAO), Rome. Available at: http://www.fao.org/climate-smart-agriculture-sourcebook/en/ (accessed 23 March 2022).

FAO (2018) *The State of the World's Forests – Forest Pathways to Sustainable Development*. Food and Agriculture Organization of the United Nations (FAO), Rome. Available at: http://www.fao.org/documents/card/en/c/I9535EN/ (accessed 23 March 2022).

IPCC (2007) *Climate Change 2007: Impacts, Adaptation and Vulnerability* (edited by Parry, M.L., Canziani, O.F., Palutikof, J.P., van der Linden, P.J. and Hanson, C.E.). Contribution of Working Group II to the Fourth Assessment Report of the Intergovernmental Panel on Climate Change. Cambridge University Press, Cambridge, UK

Jiddawi, N.S. and Öhman, M.C. (2002) Marine fisheries in Tanzania. *Ambio* 31(7/8), 518–527.

Jones, A.D., Shrinivas, A. and Bezner-Kerr, R. (2014) Farm production diversity is associated with greater household dietary diversity in Malawi: findings from nationally representative data. *Food Policy* 46(C), 1–12.

Mehar, M., Mekkawy, W., McDougall, C. and Benzie, J.A.H. (2020) A systematic literature review of fish trait preferences by end users: implications for fish breeding. *Reviews in Aquaculture* 2020, 12(3), 1273–96. DOI: https://onlinelibrary.wiley.com/doi/abs/10.1111/raq.12382.

Myers, S.S., Smith, M.R., Guth, S., Golden, C.D. *et al.* (2017) Climate change and global food systems: potential impacts on food security and undernutrition. *Annual Review of Public Health* 38(1), 259–277.

Nelson, G., Cai, Z., Hassan, R., Godfray, C., Santos, M. and Hema, S. (2012) *Food Security and Climate Change.* A Report by the High Level Panel of Experts on Food Security and Nutrition of the Committee on World Food Security. IFPRI, Washington, DC.

Opolot, H.N., Isubikalu, P., Obaa, B.B. and Ebanyat, P. (2018) Influence of university entrepreneurship training on farmers' competences for improved productivity and market access in Uganda. *Cogent Food and Agriculture* 4(1), 1469211. DOI: 10.1080/23311932.2018.1469211.

Phalkey, R.K., Arnda-Jan, C., Marx, S., Hofle, B. and Sauerborn, R. (2015) Systematic review of current efforts to quantify the impacts of climate change on undernutrition. *Proceedings of the National Academy of Sciences* 112(33), E4522–E4529.

Teklewold, H., Gebrehiwot, T. and Bezabih, M. (2019) Climate smart agricultural practices and gender differentiated nutrition outcome: an empirical evidence from Ethiopia. *World Development* 122, 38–53.

Tirado, M.C., Cohen, M., Aberman, N., Meerman, J. and Thompson, B. (2010) Addressing the challenges of climate change and biofuel production for food and nutrition security. *Food Research International* 43, 1729–1744.

Torri, C.M. and Martinez, A. (2014) Women's empowerment and micro-entrepreneurship in India: constructing a new development paradigm? *Progress in Development Studies* 14(1), 31–48.

Yadav, S.S. and Lal, R. (2018) Vulnerability of women to climate change in arid and semi-arid regions: the case of India and South Asia. *Journal of Arid Environments* 149, 4–17.

11 Gender Analysis for Climate-smart Agriculture: Perspectives and Research Implications

Abstract

How can we do gender analysis in relation to climate-smart intervention? To address this and related questions for climate-resilient strategies, we need careful research design. The present chapter elaborates on seven different steps needed for gender analysis. These steps include: understanding the focus and context of study; discipline of investigation; level of investigation; unit of analysis; effectiveness indicators; and methods and tools needed for gender analysis.

11.1 Introduction

> 'CSA [climate-smart agriculture] strategies are unlikely to be effective, let alone equitable or transformative, without active attention to gender.'
>
> (Bernier *et al.*, 2015, quoted in World Bank *et al.*, 2015)

The argument is well received by many of the researchers and policymakers including non-gender specialists or experts. In consequence, there is an increasing number of researchers which consider gender aspects, though with varying degrees of scale and stage of gender integration. Several guidelines and resource materials are prepared and available in the public domain to increase gender understanding among the researchers (see Reference Guides section). Over the years there has been a shift from the concept of the male being viewed as the head of the household or the breadwinner and women having invisible roles as supporter of sex-disaggregated data, towards gender-disaggregated data and a social relations framework. These new ways of analysing gender have nuanced ways to identify women and men, for instance *vis-a-vis* the intersections of age, class, ethnicity, etc. With the development of collective household models for decision making, there is a more promising, true and clear picture of men and women as the researchers look for improvements in methods in areas of conceptual slippages

© CAB International 2022. *Climate Change, Adaptation and Gender: Policy, Practice and Methodological Underpinnings* (M. Mehar and N. Prasad)
DOI: 10.1079/9781789249910.0011

and ambiguities in gender-related terms and contradictions related to social realities. In Chapter 1 we discussed the buzzwords as well as evolution of 'gender' in agriculture. In this chapter we focus on understanding the gender analysis aspect in the research process as visualized in Fig. 11.1.

We elaborate on seven important steps for gender research analysis. These are:

- Decide the focus and context of study. Identifying specific context and dimension will help to decide whether gender-integrated or gender-strategic research is needed.
- Discipline of investigation. Relevance to the research objective and context needs gender experts to work together with other disciplines for CSA adoption, scaling and impact.
- Level of investigation. This will help to plan for scaling and policy planning in the future.
- Unit of analysis. 'Who is to be interviewed' is a crucial question and a member of a particular household can be viewed from different angles.
- Effectiveness indicator. What key indicators ensure the effectiveness for tracking and measuring the end outcome of the programme?
- Methods and tools to measure. In addition to the previous steps, the respective methods and tools are also influenced by timing and budget available for the research.
- Gendered variables. What information is needed for gender analysis or what variables are needed for measuring effectiveness indicators are the last, but important, steps for gender analysis.

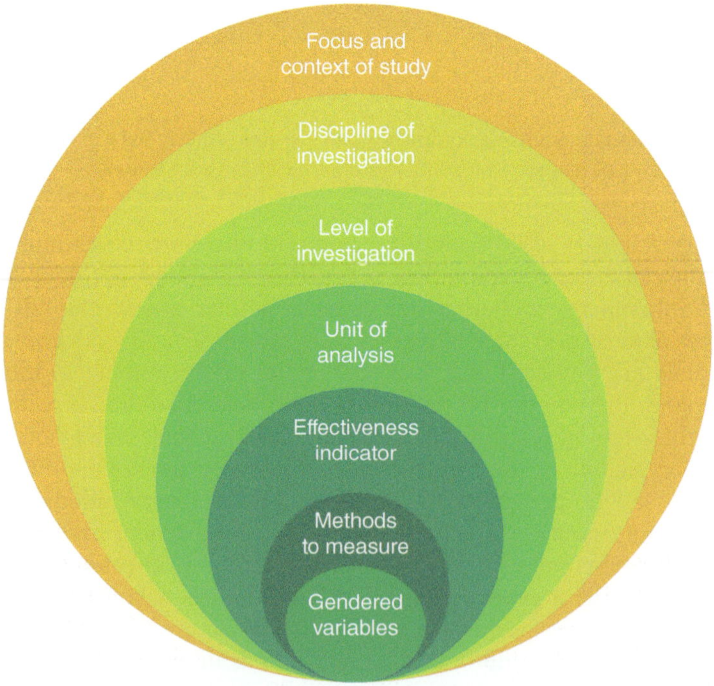

Fig. 11.1. Steps for gender research framings. (From: the author)

All these steps are sequenced and can each be imagined as a subset of previous ones, like in a Venn diagram (Fig. 11.1). We explain these in detail in this chapter, except 'gendered variable', which is discussed in the next chapter.

11.2 Context of Study and Discipline of Investigation

From the discussion in the previous chapter, we found four thematic areas important for gender analysis in terms of climate-resilient strategies:

1. Social and gender norms – context analysis: includes exploring norms, gender division of labour, which influence the access and control over inputs, knowledge, social networks and key assets and thus act as a key enabler or barrier for technology adoption.
2. Gender-responsive and -inclusive technology: includes identifying the existing technologies used by rural men and women, understanding women's and men's respective technology needs and preferences, identifying gaps in assessing and using climate-smart interventions, estimating the relevance and impact for new technology in the local socio-economic and cultural milieu, and policy environment.
3. Target beneficiaries (quality and extent of participation): includes assessment of targeted beneficiaries with a comparison of before and after technology adoption, and also includes comparing beneficiaries with non-beneficiaries.
4. Impact of technology: includes impact on climate mitigation (in terms of a reduction in greenhouse gas (GHG) emissions as well as a reduction in the use of practices leading to more GHG). This also includes tangible impacts (in terms of production, income, health, etc.) and intangible impacts (in terms of social welfare such as increasing empowerment of women).

Most of the studies (except review papers) are carried out or planned to be carried out in regions where technology dissemination programmes exist or are ongoing. In a few of the impact-assessment studies, information is collected from a control group (or non-beneficiary) to estimate the impact.

Although all social scientists try to select a perfect research design for gender analysis, we know that everyone has their own limits (including budget and time constraints) and often the research decisions are influenced by the researcher's education and training. This is outside of our control. The other limitation is the proclivities of the readers.

In previous sections we have shown a few examples directed towards increasing efforts and international agreements working for climate solutions by integrating people from different disciplines rather than just one discipline. Multi-, inter- or transdisciplinary approaches are used in climate-smart intervention programmes. Specific to gender, this, in a way, is promising as gender in such programmes can now be used explicitly from the beginning of the planning to the outcome to achieve a gender equity goal. In the interdisciplinary approach, the social scientist is expected to work with other stakeholders including engineers, breeders and extension workers to prepare a customized product relevant to different segments of society. Hellin *et al.* (2020) discuss the transdisciplinary network and co-operative efforts in the context of innovation systems and farmers' access to and use of CSA. For example, the authors discussed establishment of a Direct Seeded Rice Consortium, public–private multi-stakeholder research for a development platform which aims to address complex research issues and helps in scaling of direct seeded rice in Asia.

11.3 Level of Investigation

Gender, in the lens of the social science and climate change arena, sustains itself as a reflexive science in each level of analysis. The particular level of inquiry explains the purpose of gender at various levels with its possibilities and reason for inclusion, its procedure and limits, as well as the effects of its activity on climate change.

Methodologically, the level of investigation has three general levels: macro, micro and meso, which are relative to each other, and conceptualizing one has implications for the others. The distinct factors are agents, ideological structure, location, and size and scale of a research target. The agents vary from macro (nations, multinational bodies, institutions, social structures) to meso (communities, groups, value chain node or value chain) to micro (small groups, household, intra-household, individual) level.[1] The other distinguishing factors are differences in ideological structure divisions between macro-level perspectives on collective values (influenced by social, cultural and political norms), the meso level of intra-community networks (influenced by social identities such as agency, power, etc.) and the micro level of individuals engaging with others (influenced by individual cognition, identities, actions and motives). These ideologies are the foundation of differentiation and stratification between men and women in society. Colfer (2013) has introduced these three scales, described as the '3 Ms' – micro, meso and macro – by a gender box framework (Fig. 11.2). The framework covers 11 issues affecting women's participation and decision making in forestry management.

Two important points to consider for the level of analysis are: (i) different levels of analysis may need different variables and research methods to capture the effect; and (ii) individuals at the same time may belong to more than one level depending on how they feel and what their activities are. In consequence, in certain cases, there is a need to recognize and adopt multi-layered analysis. The three important analytical approaches to multi-layered analysis are multi-level analysis, social network analysis and discourse analysis.

Multi-level analysis involves 'modelling that decomposes the variability into at least two levels of aggregation' (Scott and Lewis, 2017, p. 1606). For example, there could be different layers for understanding the impact of adoption of CSA by women on their farm yield and food. These are: (a) whether women adopted CSA (yes/no); (b) to what extent (or on what area) they adopted CSA, as they may not have enough resources required to implement it; (c) whether it results in increasing production (yes/no); and, if it does, whether they have control over the production or income from it; and (d) how they use the production or income for their family welfare.

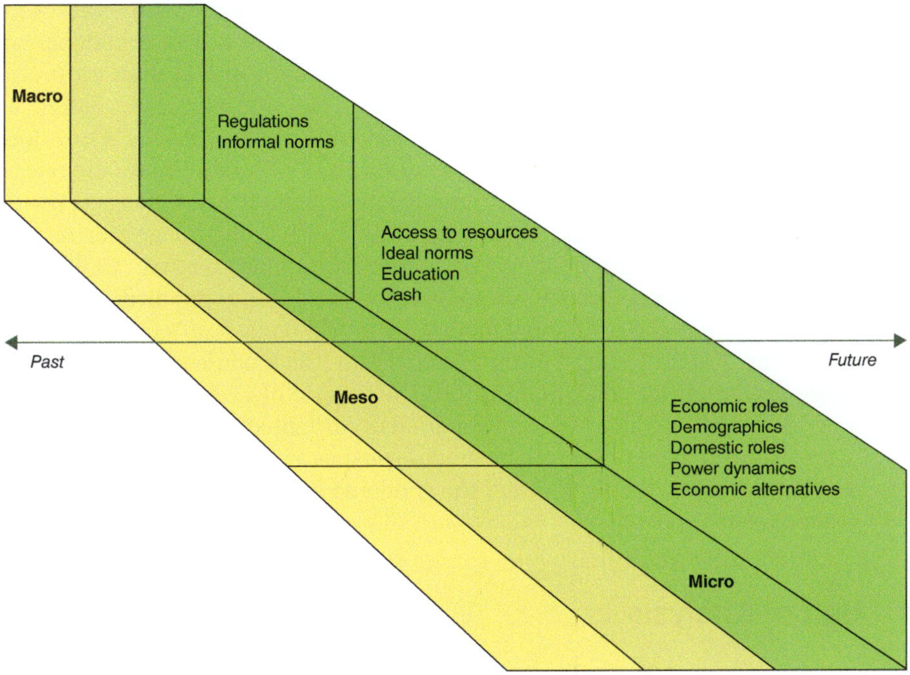

Fig. 11.2. Scale of investigation. (Adapted with permission from Colfer, 2013)

Social network analysis, by definition, 'focuses on patterns of relationships (e.g. communication frequency, advice-seeking) among nodes (e.g. team members) and offers distinctive analytical techniques for doing so' (Scott and Lewis, 2017, p. 1607). For example, Magnan *et al.* (2015) found that women in India who learn about laser land levelling (LLL) through their networks have an immediate effect on household demand for LLL. The authors have used agricultural information contacts to define social networks and assume a given farmer receives an LLL network 'treatment' if s/he has at least one first-generation adopter. LLL, a resource-conserving technology, is expensive and farmers have access to it by service providers on rental arrangements. Given the high demand for LLL, farmers with their strong networks are more likely to access it. Another example is by Villanueva *et al.* (2016) in their meso-level survey on the connections between experts (who provide climate-smart information and technologies) and farmers from the Rakai and Lushoto regions in Africa. The authors found that women in the studied site generally have smaller networks with respect to farming techniques and practices, have fewer connections to farmers or experts with whom they discuss agricultural issues, and are less likely than men to attend a farmer field day or a training workshop or to receive advice from extension officers. Natcher *et al.* (2020), using social network analysis, has examined the relative position that male and female investigators occupy in ArcticNet.[2] The authors found that female investigators are under-represented and have fewer research collaborators as compared to their male colleagues. The authors also found that male investigators tend to publish predominantly with other females as compared to females who have heterophilious collaborations.

Discourse analysis, by definition, 'conceptualizes and operationalizes micro, meso and macro levels of analysis by examining how conservations, narratives, rhetoric and tropes constitute organizing from the bottom-up processes' (Scott and Lewis, 2017, p. 1608) or 'scaling up' (Putnam and Cooren, 2004, p. 324). Good references for gendered discourse analysis are by Arora-Jonsson (2011) and Collins (2017). Arora-Jonsson (2011) discusses how 'the relegation of gender mainly to vulnerability and partly to virtuousness' detracts the attention of policymakers and institutions at all levels from the 'gender and power inequalities in decision making in environmental management'. Collins (2017) critically discusses the gender dimensions of agricultural practices in the Gender and Climate Smart Technology module of World Bank *et al.*'s (2015) *Gender and Agriculture Sourcebook*. The authors argue that 'the module makes strides toward more gender-aware policymaking, but the version of CSA discussed in the module sidesteps the market-led and productivity-oriented practices often associated with CSA. As a result, though the module pushes a more feminist agenda in many respects, it does not fully consider the gendered implications of corporate-led and trade-driven CSA.'

11.4 Unit of Analysis

The term 'level of investigation' is distinct from 'unit of analysis' in that the former refers to a more or less integrated set of relationships. The particular level of investigation or analysis shapes the research design about the topic, how and what information is to be acquired and, most importantly, from whom to gather the information. The 'whom' or 'who to interview' is the unit of analysis in social science research. An example of a unit of analysis for macro-level studies is described in Chapter 3 referring to women holding different positions in science. 'Who' are the participants of the climate policy. The meso-level 'who' could again be viewed as the gender of the person in the structure hierarchy, practices and policies, including the gender of extension workers. The agricultural extension workforce employed is male-dominated in both developing and developed countries and that is also viewed as a barrier to a female farm practice, especially when it is about adoption of technology promoted by extension workers. In a cross-country study of 97 countries in 2011, only 15% of extension personnel were found to be female (FAO, 1993).

Micro-level units, which refers to individual persons and interactive processes, are implicitly connected to micro- and macro-level structures. Depending on the level of analysis, the information needed, the methods for data collection and analysis vary. Consequently, different questions need to be asked (Table 11.1).

The analysis at household level is based on the model of the rational individual (as the member of a household) who aims to maximize his/her utility and household welfare. As discussed previously, each household member, based on age, gender and power within the home, has different roles and different access to resources as well as different amounts of time to contribute. Nepal *et al.* (2005) has illustrated the activity set and consumption bundles for household members in the simplest way (Fig. 11.3). In the figure, an individual member's total time

Table 11.1. Example of unit and level of analysis. (Revised from Mehar and McDougall, 2018)

Level of analysis	Unit of analysis	Focus	Examples of questions relating to gender at different scales, such as fish nutrition
Micro	Individual	Intra-household dynamics	What is the reason you eat fish? (circle relevant option; multiple option possible): tasty/cheap/nutritious/regular meal/other (specify).
	Household	Intra-household dynamics	Do all male and female family members eat an equal share of fish when it is cooked? (yes/no) If no, does it vary with gender, age or seniority?
Meso	Community	Inter-household dynamics	Does it happen in the same way for all families within the community? (yes/no) If no, discuss how and why families do it differently, and what they do (emphasize the role of gender, class, ethnicity, etc.).
Meso/macro	Social structure	Inter-state dynamics	Is nutrition intake from fish the same for men and women? (yes/no) If no, which states lag behind in gender-disparity malnutrition indicators?
Macro	Social structure	Society	What are the general norms of taking fish? Is it differentiated by gender and age of the household member?

can be allocated to the vector of independent activities I (1,, ..., i | I) or to the vector of joint activities J (1,..., ..., j | J). Similarly, the household consumption can be either private consumption bundle K (1,..., ..., k | K) or the shared consumption bundle S (1,..., ..., s | S). Here, I is the universal set of independent activities and K is the universal set of private consumption bundle for all household members.

The theoretical basis for understanding the intra-household decision-making process of resource allocation and consumption from an economic point of view can be viewed from the distinct development in decision-making models of Fig. 11.4. The framework of unitary and collective household models is provided by many scholars (Mattila-Wiro, 1999). The additional layer of recent thoughts is compiled by authors of this book from their reading of recent changes in decision-making models.

The common and neoclassical approach used as a unit of analysis is heads of households based on the unitary household model. For many decades, households were assumed to behave as one decision-making unit with one decision maker who acts for the benefit of all members. This approach has, however, been found to be unreliable due to limitations related to welfare and methodological estimation (Budlender, 2003). The criticism of using female-headed household (FHH) and male-headed household (MHH) comparisons are not considered appropriate as FHH are relatively poorer than MHH. Quisumbing's (2003) literature review on household decision making based on empirical studies from several countries summarizes that households do not act as a unitary model when making

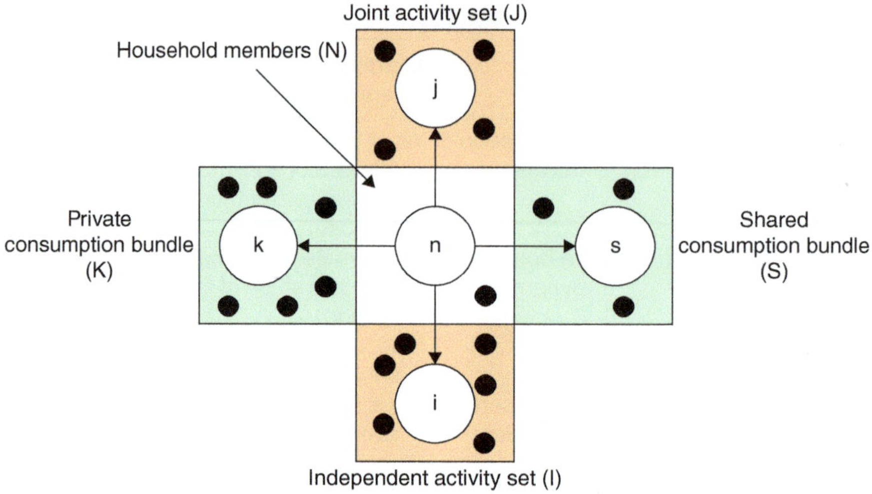

Fig. 11.3. Activity set and consumption bundles for household members. (Adapted with permission from Nepal *et al.*, 2005)

decisions. Also, this approach restricts the individual preferences or roles within the household. To consider individual roles, preferences of household members of non-unitary households or collective models are used. It is further sub-categorized based on co-operation between household members. The first is bargaining approaches, which has the implicit assumption of disparities in power and control over resources within households by gender, age and other socio-economic characteristics. The second approach does not require explicitly understanding bargaining and is based on the sharing rule, i.e. available resources are divided and distributed to each household member. The bargaining approach is explicitly different to the sharing rule in the sense that the decision-making process within a household is explicitly specified, as the model influences the 'who' for not only taking decisions but also taking action and thus implicitly devoting his/her time to the activity. Recently, the household model has been revised from the point of view of considering time allocation in the model. Coined by Becker, the time allocation model extends the collective framework and argues to consider males' and females' time for household production functions. Becker's time allocation and household production framework readily extend to this collective framework, which permits analysis of many policies and relevant issues. Beyond an individual's consumption–leisure trade-offs for time, whether males' and females' time is a perfect substitute or complement can influence the optimal level of household production and income.

The other thoughts challenge the assumption that all men and all women are the same by considering intersectionality. Gender analysis with the intersectionality approach considers gender as intersecting or overlapping with other variables like age, education, marital status, class, etc. The other prevailing thought is consideration of different generations, believing that the social norms have repercussions on a generation. For example, in climate-extreme situations, boys and girls are treated differently. Dropouts from school and a reduction in nutrition for

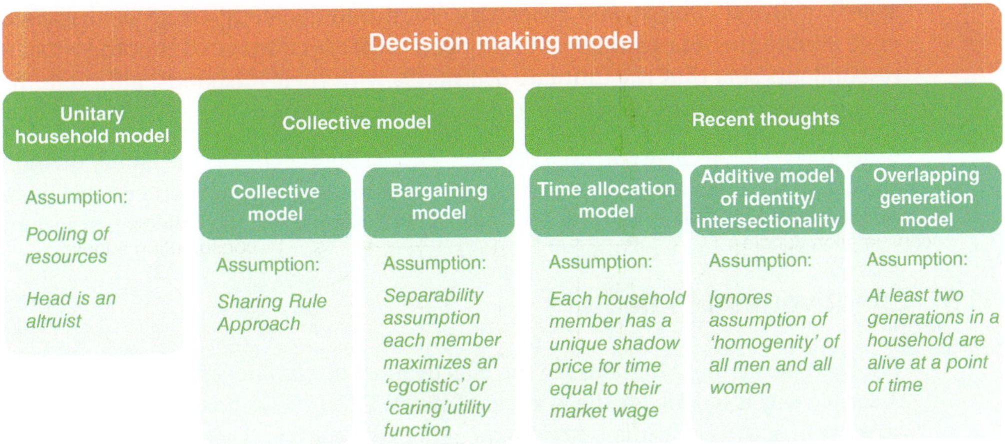

Fig. 11.4. Decision-making model within a household. (From: the author)

coping with climate change are biased towards girls as they have to help the adult women collect water and wood for fuel.

Mehar (2016) discuss the different units of analysis of gender, defined from the perspective of family relationships (head of household, husband and wife, decision-making power and control over resources) and communities (participation in organizations and/or groups). The authors have articulated the strengths and weaknesses embedded in each unit using a critical literature review (ibid.). The identity of 'who to interview' is important as different identities may have different results (refer to examples cited in Chapter 8).

In choosing and applying 'level' and 'unit' of analysis, some considerations for the researcher are:

1. There is no golden rule to justify who to interview, but there are methodological discussions and debates that can be used to decide what is relevant for the context and respective objective (Doss and Kieran, 2014; Tickamyer and Sexsmith, 2019; Tickamyer, 2020; Mehar, 2016).
2. The studies mainly focused on 'household' decisions, but it is important to look at gender unit in different nodes of the value chain, other than farmers and consumers. For example, Mehar *et al.* (2019) discuss trait preferences (though not focused on climate change) for fish by both men and women throughout the value chain from input to supply and value addition, such as hatchery and nursery operators, farmers, fish traders (processors, wholesalers, retailers, local market sellers, market experts) and consumers.
3. Within a household we cannot ignore that farming is a family affair. So all decisions are not mutually exclusive and are in the domain of relations and power. The idea is not to separate women and men but to empower both and train them to adopt improved technology for improved yield and better livelihoods.
4. In consequence, there is the concern of 'individual' versus 'joint' decisions within a household in addressing ownership and decision making (for details refer to Acosta *et al.*, 2019; Hindin, 2004). They are both complex and their meanings may also differ depending on the structure of the survey (how the question is

asked), cultural context and reflexivity. 'Joint management' could be disguised male dominance. For example, Van Koppen (2001) documented that the women respondents said that the ultimate decision is with their husbands, although 'he seldom takes a decision contrary to the consensus reached'.

Other methodological arguments persist, blending approaches in the binary division of men and women biased by social power, and also considering whether women are to be viewed as agents or victims, the later due to marginalization of women in society.

11.5 Effectiveness Indicators

The next step is how to measure the effectiveness of climate-smart interventions. A simple way is to look for the number of users. FAO (2017) has also listed gender-sensitive indicators for climate-smart agriculture awareness and adaptation (Box 11.1). Numeric indicators, as in examples 1, 2 and 3 can be calculated simply and are relevant for project success. The other variables, like 4, 5 and 6, are shaped by gender and power dynamics and need clarity about aspects like 'What is access?', 'How is poverty defined?' and 'What does control mean?' Different evaluators may have different ideas to define these. There is no single assessment framework that can

Box 11.1. Selected gender-sensitive indicators for climate-smart agriculture. (Adapted from FAO, 2017. Originally adapted from World Bank *et al.*, 2015)

1. The number of gender-responsive technologies applicable for climate-smart agriculture.
2. The rate of participation of men and women in the selection process for a climate-smart practice or technology.
3. The number or percentage of women and men participating in climate-smart agriculture-related Farmer Field Schools or farmer-to-farmer extension services.
4. The proportion of services dealing with climate-smart agriculture, including credit services, that are accessible to both women and men.
5. The number of women farmers and men farmers who have regular access to weather and climate information services and price information services, and make use of them.
6. Changes in property (e.g. land, livestock, trees) owned and controlled by women and men in different age and ethnic groups.
7. Perceptions of women and men on the usefulness of the climate-smart agriculture technology and the benefits that would accrue from its adoption.
8. Percentage change in crop yield per hectare and year as a result of the climate-smart agriculture intervention, with figures disaggregated by female-headed households, male-headed households and household members.
9. The number of farmers participating in functional associations as a result of the project, with the numbers disaggregated by sex and by type of association (e.g. market co-operative or producer association).
10. Farmers who consider themselves better off (e.g. in terms of livelihood, income, nutrition, wellbeing, social status or empowerment) due to the climate-smart agriculture intervention, with the numbers disaggregated by sex.
11. The number of women in leadership and decision-making roles or positions in the community.
12. The changes in the labour burden of women and men (e.g. number of persons reporting a significant reduction in the time spent collecting water or fuel).

adequately capture all dimensions of gendered social change processes and unpack the nature of gender and social inequalities (Batliwala and Pittman, 2010).

To sum up, the effectiveness indicators could be assessed by estimating:

1. the number of users (beneficiaries); and
2. the impact of climate-smart interventions:
   a) impact on GHG;
   b) tangible impacts: this could be measured in terms of optimal production (with efficient use of inputs), income, nutrition, consumption, health. In the long run, the impact could also be viewed in terms of indirect benefit (such as reducing workload, increasing education enrolment, improving livelihood);
   c) intangible impact: assessing empowerment and gender-transformative change.

One of the most important instructions mentioned at different points in this book is to include 'gender' from the beginning of the planning to the outcome to achieve a gender equity goal. At the planning stage, gender indicators are developed based on the research objective with the aim to track, measure and monitor the degree of gender-inclusion in programmes. In the past few years, several guidelines and manuals have been developed and are widely available to strengthen a gender-responsive approach and deliver high-quality outcomes for each effectiveness indicator. Thematic Note 3 of the *Gender in Agriculture Sourcebook* explicitly focuses on monitoring and evaluating gender through the CSA project cycle. This thematic note, 'provides a more complete picture of a project's issues, progress, outcomes, and impacts' with 'concrete examples of how to improve the monitoring and evaluation (M&E) of gender in CSA projects throughout the project cycle at the field level' (World Bank *et al.*, 2015, p. 31). Theis and Meinzen-Dick (2016) have developed a framework to distinguish between reaching women versus benefiting women versus women's empowerment.

For overall assessment of the programme aiming for the gender-responsive adaptation planning process, multi-criteria analysis (MCA) can be adapted (Table 11.2). The manual from USAID (2013) provides a useful overview of how the tool can be used in the prioritization of adaptation plans.

11.6 Methods and Tools for Gender Analysis

The selection of effectiveness indicators decides which methods are needed for gender analysis. Gender analysis is the starting point for gender mainstreaming.

Table 11.2. Example of an MCA framework for gender-responsive adaptation planning.

Adaptation options	Considered gender equality? (yes/no)	Cost (in Rs.)	Provision of separate budget for gender in planning? (yes/no)	Technically and practically feasible? (yes/no)

The methods and tools (Box 11.2) of a gender analysis can vary and be done in different ways depending on the context and research inquiry.

A prerequisite of data required for gender analysis and gender-responsive interventions is that it has to be sex-disaggregated. Table 11.3 provides a synthesis of different methods used for research, data collection and analytical approaches for gender analysis. Also, depending on the method, information can be collected from different sources, using various methods and tools.

Depending on the nature of the query, the five types of research methods mostly used are:

Qualitative methods. These methods are semi-structured and help to describe variations in phenomena, situation, issues, etc. The popular data-collection methods used here are focus group discussions, unstructured/semi-structured or in-depth interviews, observations, deep dialogues, oral/life histories, and recording behaviour. The data collected, often small in size, are predominantly in text but also there is scope for collecting some statistical data. As inferred from Table 11.3, the method is used for understanding context and gender norms, and most often used to find a reason *why* there exists gender inequality or gender differences in perception and adoption of technology, etc. Content analysis remains dominant as an analytical approach for gender analysis and is found to be done manually in Excel or by using statistical software like SPSS or qualitative software like NVIVO.

Quantitative methods. These methods are used to quantify the extent of variation in phenomena, situations, issues, etc. These are structured and based on predetermined methodology and often carried out after pilot examinations. The dominant data-collection method is a household survey collecting information from male and female members of households. We found that the majority of gender analysis related to CSA intervention has used gender as a binary variable and presented the results in simple statistical techniques (percentages, average, mode, variance) and regression (using sex of respondent as a dummy variable). In recent years, gender analysis has been carried out with intersectionality. Intersectionality assumes a woman (man) is not merely a woman (man) but always has certain socio-economic characteristics such as age, ethnicity, self-confidence, etc. This feature is captured using interactional variables in the analysis. We found that most of the qualitative methods studies have been done using different statistical and econometric software due to the large scale of the data. Although cross-sectional survey-based research is dominating in gender and climate change, there has been very little experimental/behavioural/RCT research or research based on panel data.

Box 11.2. Difference between research methods and tools.

'Methods are used for research, whereas tools are specific instruments or techniques used for collecting information' (Mehar and McDougall, 2018). For instance, for the objective to understand variation in strategies adopted by men and women, quantitative methods can be used and information (about number of strategies, ranking of strategies, etc.) can be collected from a household survey using tools (such as ranking/rating/scoring, proportionate pilling, experimental games, etc.). However, for the objective to understand the reasons in variation, information can be compiled using qualitative methods from in-depth interviews or focus group discussions in the surveyed region using tools such as ethnographic tools, capacity and vulnerability framework, resource mapping, etc.

Table 11.3. Examples of gender analysis.

Research methods	Authors	Context	Data-collection methods	Unit	Sample size	Analytical approaches
Quantitative method	Adzawla *et al.* (2019)	Adoption intensity of climate adaptation strategies	Household survey	Household member who operated his/her own farm and family head who operated own farm	300 individuals (60% male, 40% female)	Descriptive statistics, *t*-test, ANOVA
	Mehar *et al.* (2016)	Gender-differentiated coping strategies for climate shocks	Household survey	Gender of member who made decision for adopting coping strategies and who took action	Total decision makers: male (84%), female (9%), rest joint	Descriptive, multivariate probit model
	Muriithi *et al.* (2018)	Complementary and substitutability of strategies, factors influencing adaptation strategies	Household survey with plot-level information	Male-, female- and joint-managed plot	711 farming households (52% female-, 25% male- and 23% joint-managed plot)	Econometric model
	Acosta *et al.* (2019)	This research aims to address a puzzle that Alston (2014) raises: 'Why, despite the rapid adoption of gender mainstreaming across the globe, do mainstreaming processes and practices not produce greater gender equality?'	Interviews and multi-stakeholder meetings	Not specified clearly	30 expert interviews, 10 stakeholder meetings with co-authors, 16 expert interviews with ministries, 14 interviews from NGO or farmer associations	Multi-step narrative analysis which identifies story

Continued

11.8 Summary

From the above discussion, we can argue that gender analysis needs careful approaches in planning and implementation. Different forces, as discussed in the above seven steps, shape the design and outcome of the analysis. We encourage researchers to plan their outcome-orientated gender analysis during the process of research design, though there will be revisions based on information received from literature and field experiences. As a simple step, Table 11.4 is a template for use in research design and can be modified as needed.

For a performance-measured programme, a tracking table can also be prepared at the outset of research planning (Table 11.5). Evaluation at different periods can help to track as well as modify direction of achieving efficiency indicators.

Table 11.4. Planning for gender analysis.

Dimension: research theme or context	Research questions: • broad questions • specific questions • hypothesis	Identify what information is needed	Identify what information is not needed	From whom (unit of analysis) and where to collect information	Methods of collecting information	Methods of analysing information

Table 11.5. Performance measurement table.

			Actual values at different reporting period		
Outcome measures*	Baseline values	Target values	Mid-term evaluation	End-term evaluation	Differences (actual, target)

*Outcome measures can be qualitative or quantitative.

Notes

[1] There are two more levels, knowns as exosystem and chronosystem. Chronosystem is above the macro level and relates to the socio-historical changes or major events that influence people on a wider scale. Exosystem is the layer in between meso and micro.

[2] ArcticNet is a network of Centres of Excellence of Canada that brings together scientists, engineers and managers in the natural, human health and social sciences to study the impacts of climate change in the Canadian North.

References

Acosta, M., van Wessel, M., van Bommel, S., Ampaire, E.L., Twyman, J. and Jassogne, L. (2019) What does it mean to make a 'joint' decision? Unpacking intra-household decision making in agriculture: implications for policy and practice. *The Journal of Development Studies* 6, 1210–1229.

Adzawla, W., Azumah, S.B., Anani, P.Y. and Donkob, S.A. (2019) Gender perspectives of climate change adaptation in two selected districts of Ghana. *Heliyon* 5, e02854.

Alston, M. (2014) Gender mainstreaming and climate change. *Women's Studies International Forum* 47, 287–294.

Arora-Jonsson, S. (2011) Virtue and vulnerability: discourses on women, gender and climate change. *Global Environmental Change* 21(2), 744–751.

Arora-Jonsson, S. (2014) Forty years of gender research and environmental policy: Where do we stand? *Women's Studies International Forum* 47(Part B), 295–308.

Assan, E., Suvedi, M., Olabisi, L.S. and Allen, A. (2018) Coping with and adapting to climate change: a gender perspective from smallholder farming in Ghana. *Environments* 5(86), 1–19.

Batliwala, S. and Pittman, A. (2010) *Capturing Change in Women's Realities: A Critical Overview of Current Monitoring and Evaluation Frameworks and Approaches*. Association for Women's Rights in Development (AWID), Toronto, Mexico City and Capetown.

Baum, F., MacDougall, C. and Smith, D. (2006) Participatory action research. *Journal of Epidemiology & Community Health* 60(10), 854-7. DOI: 10.1136/jech.2004.028662.

Bernier, Q., Meinzen-Dick, R.S., Kristjanson, P., Haglund, E., Kovarik, C. *et al.* (2015) *Gender and Institutional Aspects of Climate Smart Agricultural Practices: Evidence from Kenya*. CCAFS Working Paper No. 79. CGIAR Research Program on Climate Change, Agriculture and Food Security (CCAFS), Copenhagen.

Brown, S.J. (2009) Spatial analysis of socioeconomic issues: gender and GIS in Nepal. *Mountain Research and Development* 23(4), 338–344.

Budlender, D. (2003) The debate about household headship social dynamics. *A Journal of African Studies* 9(2), 48–72.

Bunce, A. and Ford, J. (2015) How is adaptation, resilience, and vulnerability research engaging with gender? *Environmental Research Letters* 10, 123003.

Call, M. and Sellers, S. (2019) How does gendered vulnerability shape the adoption and impact of sustainable livelihood interventions in an era of global climate change? *Environmental Research Letters* 14, 083005.

Chambers, R. (1994) Participatory Rural Appraisal (PRA): Challenges, Potentials and Paradigm. *World Development* 22(10), 1437–1454.

Chaudhury, M., Kristjanson, P., Kyagazze, F., Naab, J.B. and Neelormi, S. (2012) *Participatory Gender-Sensitive Approaches for Addressing Key Climate Change Related Research Issues: Evidence from Bangladesh, Ghana, and Uganda*. Working Paper 19. CGIAR Research Program on Climate Change, Agriculture and Food Security (CCAFS), Copenhagen.

Colfer, C.J.P. (2013) The gender box: a framework for analysing gender roles in forest management. *Occasional Paper* 82. CIFOR, Bogor, Indonesia.

Collins, A. (2017) Saying all the right things? Gendered discourse in climate-smart agriculture. *The Journal of Peasant Studies* 45(1), 175–191.

Doss, R.C. and Kieran, C. (2014) *Standards for Collecting Sex-Disaggregated Data for Gender Analysis: A Guide for CGIAR Researchers*. CGIAR Gender and Agriculture Research Network. International Food Policy Research Institute, Washington, DC.

FAO (1993) *Agricultural Extension and Farm Women in the 1980s*. FAO, Rome.

FAO (2012) *Training Guide: Gender and Climate Change Research in Agriculture and Food Security for Rural Development*. Food and Agriculture Organization of the United Nations, Rome.

FAO (2017) *Climate Smart Agriculture Sourcebook*. Module C6 – 5 Data and indicators for gender-responsive CSA. Food and Agriculture Organization of the United Nations, Rome.

Gumucio, T., Hansen, J., Huyer, S. and van Huysen, T. (2020) Gender-responsive rural climate services: a review of the literature, *Climate and Development* 12(3), 241–254.

Hellin, J., Balié, J., Fisher, E., Kohli, A., Connor, M. *et al.* (2020) Trans-disciplinary responses to climate change: lessons from rice-based systems in Asia. *Climate* 8, 35.

Hindin, M.J. (2004) *Measuring Gender Dynamics: The Meaning of 'Joint' Decisions in Household Surveys*. Johns Hopkins Bloomberg School of Public Health Department of Population and Family Health Sciences, Baltimore, Maryland.

IFPRI (International Food Policy Research Institute) (2018) *IFPRI-CCAFS Gender and Climate Change Survey Data*. Harvard Dataverse, V1, Rakai, Uganda. Available at: https://doi.org/10.7910/DVN/CBVLK5 (accessed 25 March 2022).

Jost, C., Ferdous, N. and Spicer, T.D. (2014) *Gender and Inclusion Toolbox: Participatory Research in Climate Change and Agriculture*. CGIAR Research Program on Climate Change, Agriculture and Food Security (CCAFS), CARE International and the World Agroforestry Centre (ICRAF), Copenhagen.

Jost, C., Kyazze, F., Naab, J., Neelormi, S., Kinyangi, J. *et al.* (2016) Understanding gender dimensions of agriculture and climate change in smallholder farming communities. *Climate and Development* 8, 2, 133–144. DOI: 10.1080/17565529.2015.1050978.

Magnan, N., Spielman, D.J., Gulati, K. and Lybbert, T.J. (2015) *Information Networks Among Women and Men and the Demand for an Agricultural Technology in India*. IFPRI Discussion Paper 01411.

Mattila-Wiro, P. (1999) *Economics Theories of the Household: A Critical Review*. UNU-WIDER.

Mehar, M. (2016) Role of gender differentiated risk and time preferences in rice varietal selection in eastern India. PhD thesis, Department of Economics, School of Social Science, Indira Gandhi National Open University, India. Available at: https://shodhganga.inflibnet.ac.in/handle/10603/212284 (accessed 8 April 2022).

Mehar, M. and McDougall, C. (2018) *Methods and Tools for Gender Analysis in FISH: A Preliminary Consolidation and Reference Guide*. CGIAR Research Program on Fish Agri-Food Systems, Penang, Malaysia.

Mehar, M., Mittal, S. and Prasad, N. (2016) Farmers coping strategies for climate shock: Is it differentiated by gender? *Journal of Rural Studies* 44, 123–131.

Mehar, M., Mekkawy, W., McDougall, C. and Benzie, J. (2019) A systematic literature review of fish trait preferences by end users: implications for fish breeding. *Reviews in Aquaculture* 1–24.

Meinzen-Dick, R., Van Koppen, B., Behrman, J., Karelina, Z. and Akamandisa, V. (2012) *Putting Gender on the Map: Methods for Mapping Gendered Farm Management Systems in Sub-Saharan Africa*. IFPRI discussion paper 01153, Washington, DC.

Mudege, N.N., Mdege, N., Abidin, P.E. and Bhatasara, S. (2017) The role of gender norms in access to agricultural training in Chikwawa and Phalombe, Malawi. *Gender, Place & Culture* 24(12), 1689–1710.

Muriithi, B.W., Menale, K., Diiro, G. and Muricho, G. (2018) Does gender matter in the adoption of push-pull pest management and other sustainable agricultural practices? Evidence from Western Kenya. *Food Security* 10(2), 253–272.

Mwongera, C., Shikuku, K.M., Twyman, J., Läderach, P., Ampire, E. *et al.* (2017) Climate smart agriculture rapid appraisal (CSA-RA): a tool for prioritizing context-specific climate smart agriculture technologies. *Agricultural Systems* 151, 192–203.

Natcher, D., Bogdan, M.A., Lieverse, A. and Spiers, K. (2020) Gender and Arctic climate change science in Canada. *Palgrave Communications* 6, 32. DOI: 10.1057/s41599-020-0407-6.

Nepal, K.P., Fukuda, D. and Yai, T. (2005) Microeconomic models of intra-household activity time allocation. *Journal of the Eastern Asia Society for Transportation Studies* 6, 1637–1650.

Onta, N. and Resurrección, B.P. (2011) The role of gender and caste in climate adaptation strategies in Nepal. *Mountain Research and Development* 31(4), 351–356.

Pavlovskaya, M. and Martin, K.S. (2007) Feminism and geographic information systems: from a missing object to a mapping subject. *Geography Compass* 1(3) 583–606.

Price, R. (2018) *Women-initiated Measures to Cope with Environmental Stresses and Climate Change in South Asia.* Institute of Development Studies. Helpdesk Report.

Putnam, L. and Cooren, F. (2004) Alternative perspectives on the role of text and agency in constituting organizations. *Organization* 11, 323–333.

Quisumbing, A.R. (2003) What have we learned from research on intra-household allocation? In: Quisumbing, A.R. (ed.) *Household Decisions, Gender and Development: A Synthesis of Recent Research*. IFPRI, Washington, DC.

Rengalakshmi, R., Manjula, M. and Devaraj, M. (2018) Making climate information gender sensitive: lessons from Tamil Nadu. *Economic and Political Weekly* LIII(17), 87–95.

Salvini, G., van Paasen, A., Ligtenberg, A., Carrero, G.C. and Bregt, A.K. (2016) A role-playing game as a tool to facilitate social learning and collective action towards Climate Smart Agriculture: lessons learned from Apuí, Brazil. *Environmental Science & Policy* 63, 113–121.

Scott, C.R. and Lewis, L. (2017) *The International Encyclopedia of Organizational Communications*. Wiley-Blackwell, Malden, Massachusetts.

SERVIR-Mekong (2015) *Gender and GIS: Guidance Notes*. Asian Disaster Preparedness Center, Bangkok.

Theis, S. and Meinzen-Dick, R. (2016) *Reach, Benefit, or Empower: Clarifying Gender Strategies of Development Projects.* IFPRI Blog. International Food Policy Research Institute, Washington, DC.

Tickamyer, A.R. (2020) Feminist methods and methodology in agricultural research. In: Sachs, C., Jensen, L., Sexsmith, K. and Castellanos, P. (eds) *Gender and Agriculture Handbook*. Taylor and Francis/Routledge, New York.

Tickamyer A.R. and Sexsmith, K. (2019) How to do gender research? Feminist perspectives on gender research in agriculture. In: Sachs, C. (ed.) *Gender, Agriculture and Agrarian Transformations*. Routledge, London and New York.

Tsige, M., Synnevag, G. and Aune, J.B. (2020) Gendered constraints for adopting climate-smart agriculture amongst smallholder Ethiopian women farmers. *Scientific African* 7, e00250.

USAID (2013) *Analysing Climate Change Adaptation Options Using Multi-criteria Analysis: African and Latin American Resilience to Climate Change (ARCC).* US Agency for International Development, Washington, DC.

Van Koppen, B., Nagar, R.K. and Vasavada, S. (2001) *Gender and Irrigation in India: The Women's Irrigation Group of Jambar, South Gujarat.* Working Paper 10. International Water Management Institute, Colombo.

Villanueva, A.B., Jha, Y., Ogwal-Omara, R., Welch, E., Seyoum Wejadoo, A. *et al.* (2016) *Influence of Social Networks on the Adoption of Climate Smart Technologies in East Africa: Findings from Two Surveys and Participatory Exercises with Farmers and Local Experts. CCAFS Info Note*. CGIAR Research Program on Climate Change, Agriculture and Food Security (CCAFS), Copenhagen.

World Bank Group, FAO and IFAD (2015) Gender in climate-smart agriculture. Module 18 for *Gender in Agriculture Sourcebook*. World Bank, Food and Agriculture Organization of the United Nations, and the International Fund for Agricultural Development. Available at: https://openknowledge.worldbank.org/handle/10986/22983 (accessed 10 April 2022).

12 Mediator and Moderator Variables in Gender Analysis

Abstract

Is gender equity the cause or effect of the adoption of climate-smart intervention? In previous chapters we discussed how gender inequalities led to differences in perception and acted as enablers or barriers in the adoption of climate-smart interventions in relation to gender. For uptake and scaling of these interventions, often dissemination or demonstration programmes educate and provide necessary skills to farmers. An interesting question is whether uptake of these technologies needs a pre-prerequisite of gender equity, and also if, as a result of some programmes, there is gender equity. This chapter, with the help of mediator and moderator approaches, discusses how gender is crucial in effecting climate-smart intervention adoption and, in turn, reducing climate risk as well as increasing yield.

12.1 Introduction

The gender of farmers determines what is expected and permitted for men and women in a determined context. As narrated in Chapter 8, men and women have differences in perception, priorities and preferences, which partly affect their adoption decisions. The other factor that acts as an enabler or barrier in adoption by men and women is social and gender norms in a culture-specific context. In consequence, the examination of factors influencing adoption decision making with regard to gender becomes relevant.

The relation between gender and climate-smart intervention is assumed to be unidirectional, and gender is used as a binary variable for gender analysis. In recent additions, as discussed in Chapter 9, researchers have gone beyond the simple account of the bivariate cause-and-effect relationship. They revised methodologies to see what explains this relationship and also what alters its magnitude and direction. Various variables, which influence individuals' cognitive behaviour (as discussed in Chapter 4), as well as intersectionality variables (as discussed in Chapters 4, 8 and 9), act as mediator- and moderator-gendered variables. Mediator

DOI: 10.1079/9781789249910.0012

(Me) and moderator (Mo) are two theories which are based on the assumption that a third intermediate variable exists and operates between the two variables (cause and effect).

In this chapter, we discuss the concepts of moderator and mediator variables, and identify the relevant Mo or Me variables influencing gender for adoption of climate-smart intervention. To identify these variables, we consider an example where a climate-smart intervention programme has trained both men and women in a village about climate-smart practices, uses and benefits. After the training, whether the gender of the participants influenced the adoption of climate-smart practices became the subject of research inquiry. To address the inquiry, we explore how, why and for whom the variables, including socio-economic characteristics, institutional support, social norms, individual attitude and perception, etc., act as enablers or barriers for adoption decisions.

12.2 Mediator and Moderator – Conceptual Understanding

'Mediation[1] and moderation are two theories for refining and understanding a causal relationship' (Wu and Zumbo, 2008). Both theories are based on a unidirectional trivariate hypothesis. The theory assumes that between independent variable or cause (X) and dependent variable or effect (Y) there exists a third variable (Z). The trivariate hypothesis assumes that X and Z affect Y rather than the assumption there is an association between X, Y, Z.

Though both theories are about the mechanism through which a third variable operates between a cause and an effect, mediation ≠ moderation. Table 12.1 discusses some of their common and distinguishing features.

Though, as mentioned before, mediator and moderator are not same, in real-world scenarios there are possibilities in which both interact with each other. For example, if the same set of measures are considered first as a conditional effect (moderation) and then as a mechanism (mediation) in the cause-and-effect relationship, the process would be viewed as a moderated-mediation effect. Similarly, there is a moderated-mediation relationship (Box 12.1).

Box 12.1. Relevant references.

- Wu and Zumbo (2008) – This paper described the conceptual foundation, research design, data analysis, and inferences involved in a mediation and/or moderation investigation in both experimental and non-experimental (i.e. correlational) contexts.
- Muller *et al.* (2005) – This article discusses the process where mediator and moderator can be combined such that moderation is mediated or mediation is moderated. It also provides analytical strategies for assessing each process.
- MacKinnon (2011) – This article describes mediating variables and moderating variables with examples and statistical models. It also provides reasons for integrating them in outcome studies.
- Halls and Sammons (2013) – This chapter seeks to provide a critical up-to-date review of the terms mediation, moderation and interaction as they are being commonly defined, discriminated, used and tested as of 2013, and considers some of their implications for quantitative educational research.

Table 12.1. Mediator versus moderator. (Compiled from Muller *et al*., 2005; Wu and Zumbo, 2008; MacKinnon, 2011; Halls and Sammons, 2013)

	Mediation	Moderation
Common	Both theories aim for refining and understanding a causal relationship and are unidirectional by nature.	
Definition	A mediator is a third variable that links a cause and an effect.	A moderator is a third variable that modifies a causal effect.
Hypothesis/ type of question	The hypothesis of mediation asks, '**How** and **why** there exists a pre-established causal relationship between a dependent and independent variable?'.	The hypotheses of moderation asks, 'For **whom** and **when** does an independent variable cause an effect for a dependent variable?'.
Common terms	Also termed 'intermediate variable' or 'explanatory link', 'intervening' or 'process' variables.	Also termed as 'effect-modifier' or 'causal interaction effect'.
Role	A mediator acts in dual roles: dependent variable for X; independent variable for Y (Wu and Zumbo, 2008).	Mediator has a single role: auxiliary independent variable for Y (Wu and Zumbo, 2008.)
Graphical illustration of the hypothesis	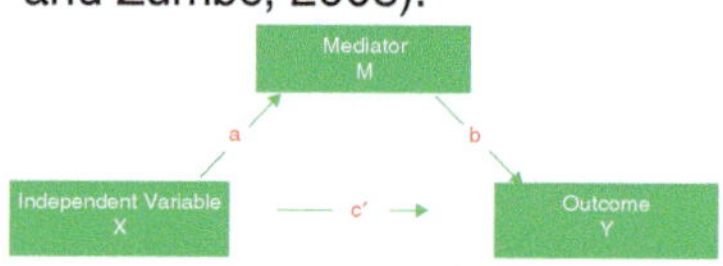 **Fig. 12.1.** Illustration of Mediation.	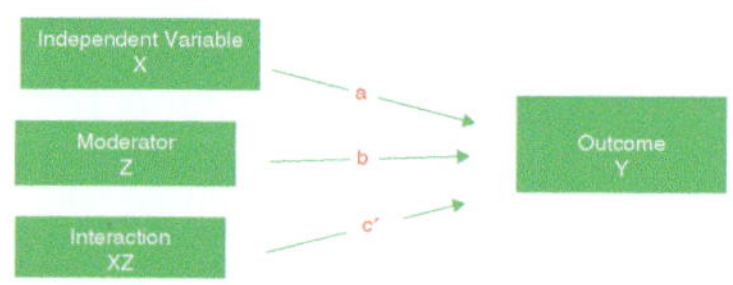 **Fig. 12.2.** llustration of Moderation.

Note: mediators have 'mediating effects' which are also referred to as 'indirect effects', 'surrogate effects', 'intermediate effects' and/or 'intervening effects'.

12.3 Gender in Climate-smart Intervention: Mediator or Moderator Variable?

As briefed in the introduction section, let us consider a setting where a climate-smart intervention programme is initiated. Under this programme, farmers, both male and female, are invited to attend training about technologies and practices that can help in reducing climate risk in adverse climate situations. The information about the programme is spread in the village by the village head and agricultural extension workers. The male and female farmer participants come of their own volition to learn after hearing about the programme. A few female farmers leave the training early after learning the training will be provided by a man, and also because they have to travel to the village school to attend the training. After the programme finishes, the researcher carries out a survey to find out who among the trainees has adopted the technologies and practices that they have learned about on their farm in the current season. The researcher found that not all the participants have adopted the practices and also found variation in the intensity of adoption. Here intensity is measured in terms of how many strategies were adopted. Another striking finding was significant differences between male and female participants' behaviour in adopting as well as intensity of adoption.

In this scenario, for estimating indicators (adoption and intensity of adoption) related to climate-smart intervention, two important questions related to the outcome of the research are: 'How does the climate-smart intervention adoption occur and why are there differences in male and female behaviour?' and 'Under what conditions and for which gender does the intervention work?'

12.3.1 Gender as a binary variable

Scholars have addressed these questions by descriptive statistics or regression models. In the former, simple percentages or frequencies are provided for male and female separately. In the regression model, gender is used as a binary dummy variable (with male=1, female=0, or otherwise). In this approach, an independent variable (gender of farmer) is assumed to have a unidirectional relation to the dependent variable (technology adoption decision, yes=1, no=0). Fig. 12.3 demonstrates gender as a binary variable using a path diagram. The direct effect of X on Y, denoted by c′.

However, this is an incomplete picture. As discussed in Chapter 4, an individual's 'cognitive' perception drives her/him to make the climate-relevant decisions for adopting or not adopting the intervention. The cognitive perspective could be measured using mediator or moderator variables, which could be qualitative, quantitative or both, depending on the research enquiry.

12.3.2 Mediator variable in gender analysis

The first question raised above is about the mediator variable. 'A mediator is often a cognitive, affective, physiological, motivational state that functions as a person's psychological process after receiving a stimulus such as intervention treatments' (Hoyle and Robinson, 2003). A mediator model is where the gender of the participant in the intervention programme (X) is hypothesized to affect their 'knowledge, skill and confidence' (Me) to adopt the strategy, and, in turn, to affect their decision to adopt (Y); this, in turn, affects farm yield. In simple terms, X----> Me---- >Y. The needs of the mediator are based on the evidence in literature that male farmers are found to be in a better position for learning as compared to female farmers. Thus gender (X) has been found to have a direct effect on the mediator, denoted as 'a'; and, in turn, the mediator has an effect on Y, denoted as 'b', as shown in the path diagram of Fig. 12.4. c′ denotes the direct (i.e. unmediated) effect of X on the outcome. The total effect is the sum of direct (c′) and indirect effects (a*b); so the total effect is c= c′+a*b).

In our basic setting, we mentioned that females leave training sessions due to social norms. This is consistent with findings in several studies (such as Tsige *et al.*, 2020; Quisumbing and Pandolfelli, 2010) which indicate that gender norms prohibit women's membership in such organizations and also put constraint on women's mobility outside their village, and, in some religions, their freedom to talk to other males outside the family. All such variables act as gender mediators and are often found to be discussed and explored using qualitative methods.

12.3.3 Moderator variable in gender analysis

The second question raised above is about moderator variables. 'A moderator is characterized as an innate attribute (i.e. ethnicity), a relatively stable trait (i.e. personality type or disposition) or a relatively unchangeable background, environmental or contextual variable (i.e. parents' educational level or neighborhood)' (Wu and Zumbo, 2008). Moderator variables influence and measure the strength of the relationship between independent variables (X) (gender of farmer) and dependent variables (Y) (say, adoption decision of resilience technologies) in three ways as depicted in Fig. 12.5. The most commonly used practice is to

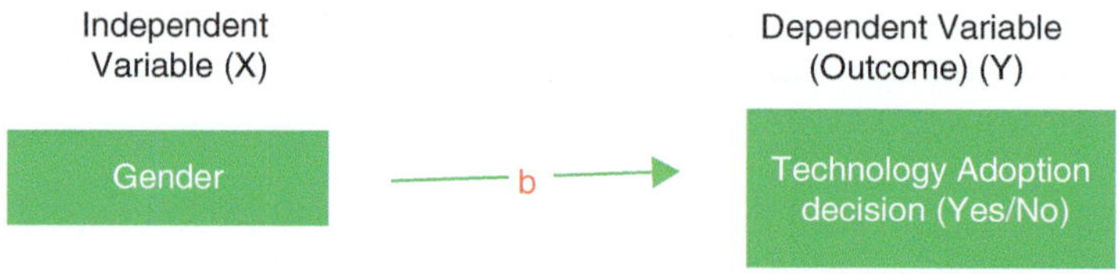

Fig. 12.3. Illustration of the binary relation.

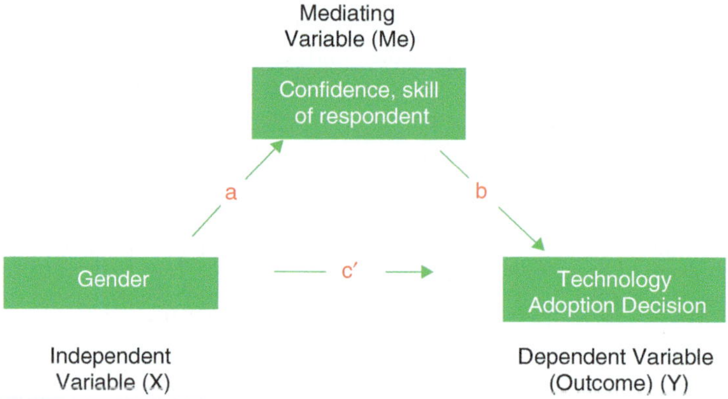

Fig. 12.4. Illustration of the mediator effect.

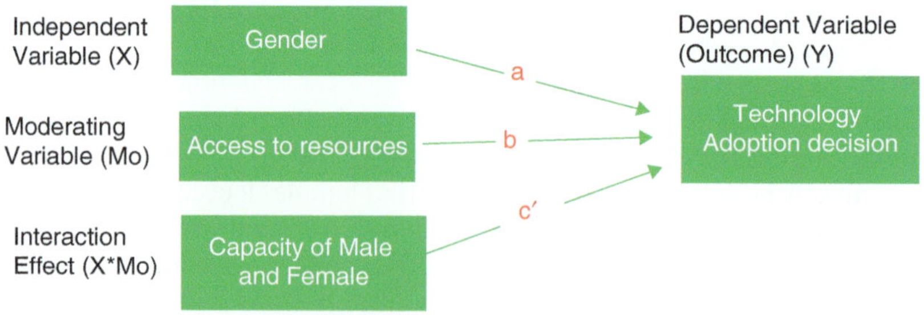

Fig. 12.5. Illustration of the moderator effect.

map the moderator variable (here adaptive capacity) using the interaction term as specified in the diagram. Adaptative capacity (Mo) measures here the farmer's access and control over resources and extension workers.

In our example, the moderator variable can be used to identify disadvantaged groups in the training. The recent efforts of carrying out gender analysis with an intersectionality approach is a good example of the moderator effect. Intersection variables such as age, class and ethnicity have a moderating effect on male and female participants' adoption decisions.

12.3.4 Mediated-moderator variable in gender analysis

The mediated-moderator model is primarily a moderator at its foundation, and the moderator has a secondary role in explaining the process of moderation effect (Wu and Zumbo, 2008). For example, when considering the adoption decision for climate strategies, we see women are less confident than men. A researcher may further investigate whether such mediation effect is consistent and has similar magnitude across genders of different adaptive capacities (Mo). This examination is also known as a mediating-moderator effect. As depicted in the path diagram in Fig. 12.6, the strength and direction of the mediation effect leads to a new path, c, in addition to path, b of moderator effect only.

This example suggests that even if males and females have equal resources, the difference in perception, self-confidence, risk-attitude, etc. can influence their decisions of intervention adoption (Y) and thus agricultural yield. Women tend to underestimate their ability relative to men, especially in settings outside their home territory. Recognizing this fact, several initiatives include programmes for women's empowerment, which influence their confidence in decision making and leadership skills. FAO and CARE (2019) have included good programmes and practices for integrating the genders and women's empowerment in climate-smart agriculture (CSA) programmes. For example, the Global Water Initiative East Africa (GWIEA) programme focuses on women's empowerment to enhance their voice within decision-making forums and institutions (ibid.).

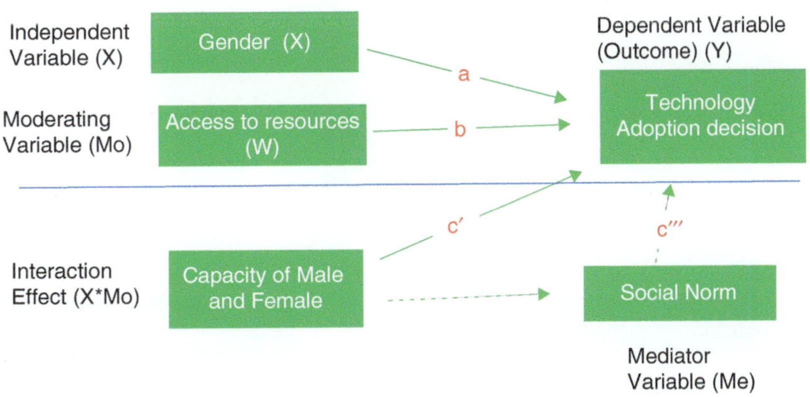

Fig. 12.6. Illustration of mediated-moderator.

12.3.5 Moderated-mediator variable in gender analysis

The moderated-mediator model is primarily mediational at its foundation, and the moderator has a secondary role in explaining the mediation effect (Wu and Zumbo, 2008). An example of gender as mediator or moderator effect would be if the gender of participants in a training programme (X) is shown to be more effective for participants with more adaptive capacity (Mo). Such a synergistic effect then leads to the likelihood of adoption of technology (Y) by the participants. The total effect is equal to: direct effect (c′+ c‴ W) + indirect effect (ab + c″ Wb).

This example suggests gender and social norms resulting in differences in mediator variable (confidence, risk perception, etc.) have multiplier effects if they have different adaptive capacity. For example, women, especially married women, in some cultures are prevented from communicating with men outside their families (Tsige *et al.*, 2020). This affects their access to extension services. In another example, women's control over cattle for the farm field is restricted and they are prohibited from ploughing. Such restrictions limit women's adaptive capacity to own oxen and use them for ploughing.

Table 12.2 provides a potential list of variables for mediator and moderator for climate-smart intervention adoption by male and female farmers. The variables are contextual and depend on which indicators are to be measured as well as which methods are used. Also, there are conceptual and statistical challenges to identify, collect and measure true mediating or moderating, or both variables. Because of the difficulty in specifying and interpreting, general research is based on a single moderator or mediator variable despite the mediator-moderator and moderator-mediator effects.

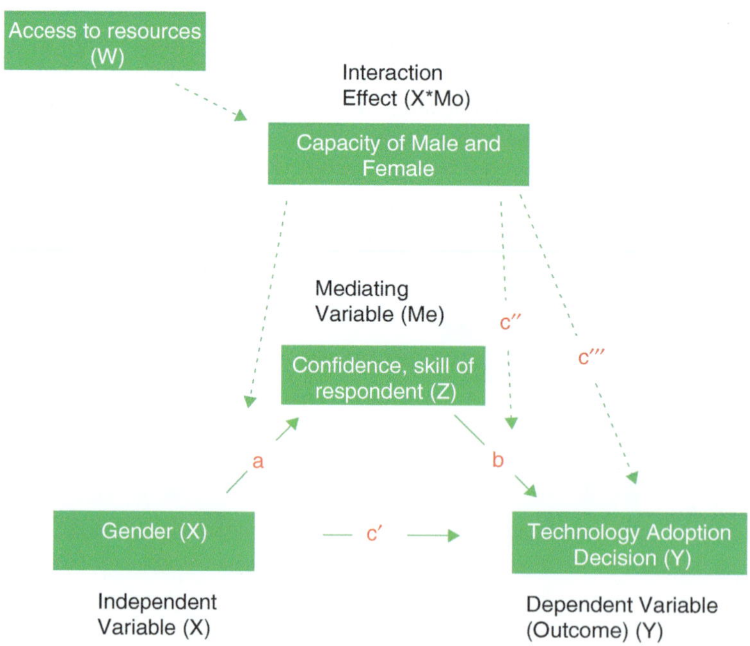

Fig. 12.7. Illustration of moderated-mediator.

Table 12.2. Potential moderator and mediator variables for gender analysis. (From: the author)

Moderator variables	Mediator variables
1. Socio-demographic characteristics: education, age, class, marital status, number of children or household size, skills 2. Adaptive capacity measured by five capitals*: i. human (skills, knowledge, and health) ii. physical (shelter, transportation, energy, communications) iii. social (networks, associations, relationships of trust, access to social institutions) iv. financial (savings, access to credit remittances, pensions) v. natural (stock of available natural resources including land, water, forests, indigenous plants, animals, and insects 3. Policy and Institutional related variables • Such as access to extension worker, access to market, access to weather information	1. Cognitive variables: • knowledge • beliefs about climate change • beliefs about risk, costs and benefits of CSA 2. Attitude towards new technology 3. Motivational variables • intention to adopt or discontinue previous adoption • social influence 4. Self-efficacy (knowledge and ability) 5. Previous event experiences

*obtained from Scoones (1998); Haglund *et al.* (2011)

Summary

Mediation and moderation methodology is mostly used in social work, health-related research, educational research, etc. We have not come across any article which establishes cause–effect in themes such as 'gender agriculture', 'gender climate change' or 'gender climate-resilient strategies', where these terms are used explicitly.

This chapter provides conceptual understanding and theoretical formulations with examples of a causal relationship between gender of the farmer and climate-resilient strategies. The discussion of mediation and moderator models is not merely of data-analytical techniques but needs to be integrated into research programmes from theoretical foundation to setting and achieving targets.

Note

[1] Mediator is different from confounder, which influences both the dependent variable and independent variable, causing a spurious association.

References

FAO and CARE (2019) *Good Practices for Integrating Gender Equality and Women's Empowerment in Climate-Smart Agriculture Programmes*. CARE USA, Atlanta, Georgia.

Haglund, E., Ndjeunga, J., Snook, L. and Pasternak, D. (2011) Dry land tree management for improved household livelihoods: farmer managed natural regeneration in Niger. *Journal of Environmental Management* 92, 1696–1705.

Hall, J. and Sammons, P. (2013) Mediation, moderation & interaction: definitions, discrimination & (some) means of testing. In Teo, T. (ed.) *Handbook of Quantitative Methods for Educational Research*. Sense Publishers, Rotterdam, The Netherlands, pp. 267–286.

Hoyle, R.H. and Robinson, J.I. (2003) Mediated and moderated effects in social psychological research: measurement, design, analysis issues. In: Sansone, C., Morf, C. and Panter, A.T. (eds) *Handbook of Methods in Social Psychology.* Sage, Thousand Oaks, California, pp. 213–233.

MacKinnon, D.P. (2011) Integrating mediators and moderators in research design. *Research on Social Work Practice* 21(6), 675–681. DOI:10.1177/1049731511414148.

Muller, D., Judd, C.M. and Yzerbyt, V.Y. (2005) When moderation is mediated and mediation is moderated. *Journal of Personality and Social Psychology* 89(6), 852–863.

Quisumbing, A.R. and Pandolfelli, L. (2010) Promising approaches to address the needs of poor female farmers: resources, constraints, and interventions. *World Development* 38(4), 581–592.

Scoones, I. (1998) *Sustainable Rural Livelihoods: A Framework for Analysis*. Institute of Development Studies, London.

Tsige, M., Synnevag, G. and Aune, J.B. (2020) Gendered constraints for adopting climate-smart agriculture amongst smallholder Ethiopian women farmers. *Scientific African* 7, e00250.

Wu, A.D. and Zumbo, B.D. (2008) Understanding and using mediators and moderators. *Social Indicators Research* 87, 367–392.

13 Methodological Applications for Future Research Plans for CRS

Abstract

In this chapter, we examine methodological issues underpinning the integration of gender mainstreaming strategy into climate change policy and implementation of adaptation strategies. A synthesis of previous chapters outlining the climate change and gender concerns is provided. For efficient gender mainstreaming in policy and actions, the key components in planning are laid out, with emphasis on implementation using the action research model.

13.1 Introduction

Often in research programmes or studies, gender is referred to as transversal[1] – cross-cutting, cross-connecting or cross-current, rather than an individual theme. This is partly a reason for why, often, gender issues are considered with an 'add and stir' approach. This leads to misrepresentation of the situation. However, the reason for referring to gender as an integrated theme is to encourage planners to:

- make an analysis of the effects of a programme on women and men, respectively, before decisions are taken and implementation processes begun; and
- systematically integrate respective priorities and needs required for participation of women as well as men throughout the project cycle.

There is a variety of issues to be considered for gender mainstreaming in the uptake and scaling of climate-resilient strategies (CRS). These include the context in which the research is conducted, activities and crop stage CRS benefits, including men's and women's roles, means of checking, warning signs, and actions needed for CRS, and identifying who should take them.

We first provide a recap of previous chapters to understand where we stand in the process of gender-inclusive CRS uptake. Later, we explore the key components

DOI: 10.1079/9781789249910.0013

needed to be considered for designing gender-mainstreaming strategies for CRS and the action research model required for implementation.

13.2 Recap of Previous Chapters

We start with key takeaways from previous chapters to identify areas of improvement and challenges for future research. A recap of the previous chapters focusing on individual entities suggests the following:

1. **Gender and CRS contextual understandings.** Understanding gender constraints and needs matters to the programme because of the following issues:
   - We have come a long way in understanding gender and climate change linkages and that is evident from the huge literature published. The continuous and increasing trend of research in this domain has positioned the gender theme well in national and international climate policies and programmes.
   - Men and women are often impacted differently by climate change; they have different reasons, different priorities and different options in response to climate issues. All these differences vary in magnitude across geographies and households depending on their household assets, social status, bargaining power within the household, etc.
   - Women are featured to be more vulnerable, which is true to a large extent. However, it is important to acknowledge that they are not only climate victims but are also active in solving them and have their own indigenous methods to address climate-related issues.
   - There is increasing concern for, and a trend of, making gender-responsive policies at international and national level. At local level, people-level outcomes and gender-related efforts are not significant.
   - Design of clientele-responsive technology and tailoring its implementation and planning according to the needs of the participants and embedded social norms in the context in which they live are important to pursue.
2. **Multiple frameworks for gender analysis.** In essence, previous chapters discussed gender and climate-smart agriculture (CSA) around the following classical gender analysis questions proposed by Hills (2011)[2]:
   - Who does what? How? Where? When? Why? (labour)
   - Who uses what? How? Where? When? Why? (access)
   - Who controls what? How? Where? When? Why? (decision making and control = power)
   - Who knows what? How? Where? When? Why? (information = power)
   - Who benefits from what? How? Where? When? Why? (benefit sharing)
   - Who is included in what? How? Where? When? Why? (participation)
3. **Approaches and concerns.** Given the depth, spread and contextual diversities of CSA and gender issues, the methods and approaches used in literature are need-specific based on the authors' judgments. There is no gold-standard rule/approach to find (feasible) ways to integrate CSA and gender. However, we argue that use of best and appropriate practice can help to understand the situation, and planning for relevant solutions is feasible in the context. Some of these practices include:

Table 13.1. Synthesis of climate change issues and gender concerns.

Climate change issues	Gender concerns
Survival security	More women than men die or are injured in climate shocks (drought, flood, etc.). Males play first responder role; search-and-rescue burden. Displacement: males migrate to risky living situations and also suffer from stress from isolation from family.
Health issues	Increased diseases and body absorption capacity.
Food and water security	Women considered natural custodians of food and water arrangements; have to bear increased burden post-climate events.
Additional burden	Domestic violence as a result of stress due to climate stress – affects women. New domestic roles. Greater female headship with immediate/temporary migration. Exploitation (low wage, long work hours, hard work); sexual and physical violence faced by men and women at workplaces.
Climate solutions	Men and women have different indigenous ways of reducing risk as well as different perception of CRS technologies.
Mechanism of adaptation strategy	Design of CRS is mostly men-friendly. Implementation processes increasing, including efforts to include both men and women. In the name of gender, dominance of women's programmes and exclusion of men.
Policies	International and national policies have long recognized the need for gender-inclusive policies. However, at local level they are not included substantially.
Barriers of adaptation	Access and control over physical assets, decision power, social norms, etc.
Gender-inclusiveness efforts	Recent efforts have used participatory approaches for including men and women. Considering women's marginalization and social concerns, efforts for increasing women's capability and making them empowered are visible. Researchers are trying to understand gender with intersectionality.

- use of different approaches including participatory to understand the context, as surveys alone may not be the best tools for capturing nuanced gender dynamics;
- making sense of information via analysing 'intersectional points' *vis-à-vis* cultural concerns (e.g. norms, behaviour, etc.) and thus highlighting the need to pay attention not only to gender but also to class and location (urban, rural, smallholders, transnational corporations);
- triangulation and mixed method: recommending the use of a different set of data to understand the complex layers – for mixed method, it is recommended to plan the timing (sequence and layering); and priority of quantitative and qualitative approaches in the research design;
- the need of focused interaction of different disciplines (multi-, inter-, trans-discipline) at planning and implementation stages.

One important point arising in the recap is that, until recently, in most of the work, gender is considered as 'add gender and stir' and often receives attention only during the data collection and analysis process. However, it should be considered in the research planning and design process at the start. Gender analysis informs gender integration at each stage of the project cycle: proposal planning and development, design, implementation and management, monitoring and evaluation (M&E), reporting, and documentation. There is increasing urgency to implement mandates to seek gender equity in climate-mitigation research, policy and practice. The failure to collect, report or analyse even the most rudimentary sex-disaggregated data, let alone more complex issues of sex and gender, is well documented and results in large gaps in understanding the shortfall in adoption of climate-resilient strategies.

13.3 Gender Mainstreaming Strategy for CRS

'Gender mainstreaming is the process of incorporating a gender perspective into any policy, action, or legislation in order to ensure that the concerns of all stakeholders are addressed and that institutional means do not perpetuate gender inequalities' (Alston, 2014). Since the World Conference on Women in Beijing in 1995, which called for gender equality and empowerment of women, gender consideration in climate policies and programmes has been considered by governments and by domestic and transnational organizations. Despite this long history, gender mainstreaming theme and vision is not clear, and has different interpretations by people of different cultures and thus has failed to bring changes to achieve gender equality. The paradox of rapid gender mainstreaming adoption and no gender equality achievement is called 'the radical potential paradox of gender mainstreaming'.

What could be an agenda, key dimensions and approaches for gender mainstreaming and equality in uptake and scaling of CRS in general and CSA in particular? To address this question, we explore the key components for gender-mainstreaming strategy for CRS:

1. Agenda. Specifying people-level planning and outcome can lead to better gender mainstreaming during project implementation. Thus, the project agenda should aim for achieving gender-related indicators with regard to the '6 As'. The '6 As' are:

Appropriate (promoting appropriate practice specific to context);
Awareness (making the users aware about technology and how to use it);
Adoption (encouraging users to adopt it by tailoring strategies at local level, overcoming key gender barriers, differences and opportunities);
Availability (making the strategies available in equitable and substantive manner in the local domain);
Affordability (ensuring that it is reasonably priced and profitable); and
Accommodative (adjustable to personal and local needs and available resources).

In addition to these 'As' for adoption and scaling of CRS, the associated benefits should also be understood, as CRS's impact is more than productivity improvement; it will also benefit the society with increased share of income, food, health benefits and other benefits. Thus, for multiplier and devastating impact, in addition to outreach (i.e. number of people using CRS), the extent and intensity (number of CRS adopted and the percentage of cultivated area where it is applied) is also important to increase. This requires gender-sensitive stakeholder consultation, sex-disaggregated data collection and gender equality training. In a few places it also requires institutional reforms and building infrastructure to support.

2. Consideration of climate policies and commitment in the research framework. One challenge identified in the recap is the gap between international or national policy formulation and understanding, and acceptance of these at local level. One way to address this issue is to include the local level concerns in policy framing and keeping the policies flexible, though with the condition of ensuring gender equity. The local organization and staff should also be informed about the policies at macro level, and there needs to be an understanding of gender-inclusive and gender sensitivity with 'do no harm' policies (Aguilar, 2016).

3. Gender-inclusion understanding. To limit the slippage from 'gender' to 'women' in programmes, clear parameters should be set in the research design (Fig. 13.1). These could be to direct clear outcomes such as:

- whether women are the only beneficiaries in the project;
- whether the target beneficiaries are both men and women and whether they will benefit equally from the project; and
- whether women, as an outcome of the programme, will be able to have a voice to discuss their opinions equally with men in the decision making.

There are two important aspects that are critical for gender-inclusive planning. First, women's voices matter, even if they are not producers. Mehar *et al.* (2019) found that women as consumers and with cooking responsibilities suggest that the producers are predominantly male, in respect of which fish species to farm. An implicit assumption is that some of the farmed fish are consumed in the family. Secondly, males and females may have individual or different, or joint or shared, preferences. This, in turn, suggests a need for different strategies.

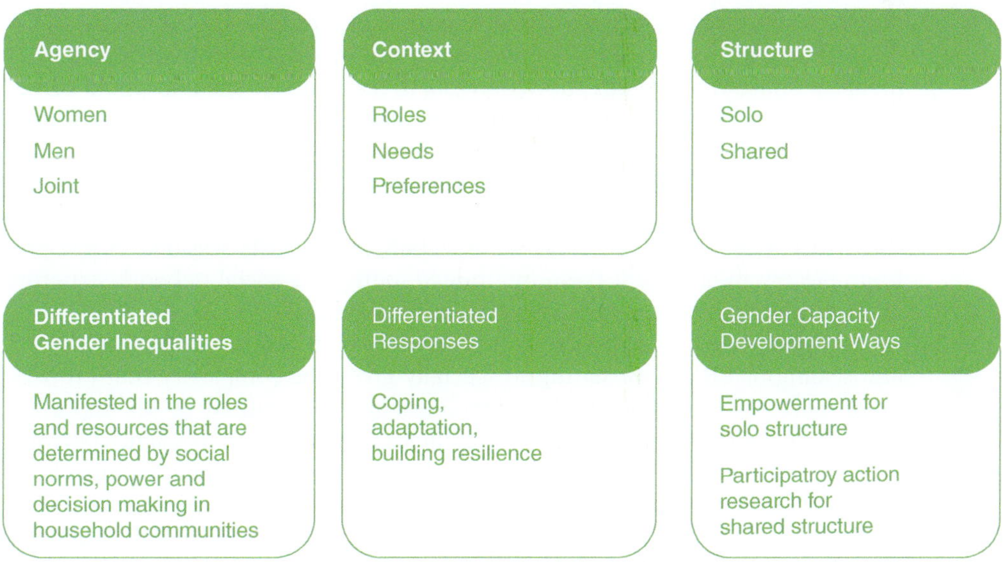

Fig. 13.1. Gender mainstreaming in the context of capacity development of targeted population.

For example, for shared preferences, social gender norms are not an issue, but for individual preferences, especially of females, strategic planning is needed, which is conducive to social norms, or else planning is needed with gender-transformative approaches.

For shared structures, participatory action research is more relevant as it creates space for women and men to critically reflect upon underlying social norms and power relations in their households and communities.

Decisions for gender integration need to be made based on concerns of both men and women: identify which technologies are most likely to be of use to and affordable for each. Have any of these had a negative effect on their labour, thus income or health? What supporting services are needed in terms of inputs, knowledge and skill to operate them and what extension services are required? How do we engage, convince and train farmers, especially women, to adopt the technology? Integrating gender into stocktaking can ensure that implementation processes are built on gender-mainstreaming capacity needs assessment, and it also helps in constructing tailored training in the implementation process. Stocktaking is the process of 'identifying available information on adaptation activities, climate change impacts, vulnerability and adaptation and assessing gaps and needs of the enabling environment for the NAP process' (UNFCCC, 2012, p. 18).

4. Situation analysis and conditions for planning stage of gender mainstreaming. This includes 'need assessments' of the targeted population in order to identify the problems and concerns of men and women in the community and then prioritize them. It is also crucial to do a situation analysis of the research context and culture to examine and identify the roots for implementation. Based on need assessment and situation analysis, the conditions include: building an implementation plan; devising a structure to execute the plan; identifying the resources available and required; level of expertise needed in the process, including gender expertise; and recognition of local stakeholders and how to involve them in the process. The need to understand context and connecting with society, collaboration with stakeholders, planning, and resources and finance is outlined below.

Connecting with society. Women are bound by social norms such as lack of mobility. Those who are not allowed to talk to men outside the household can be approached indirectly by targeting their immediate contacts such as close friends and family members. Integrating intersectional understandings of social norms in disseminating CRS can help in bringing more women into the training programme. For example, some of the intervention programmes invite women with their husbands and mothers-in-law, especially in the early sessions. A few programmes try to approach them by understanding the social network structure and identifying the strong and weak ties to the women.

Collaboration. A single entity may not be successful for implementation and scaling of CRS in all settings, especially given the complexity that prevails. Thus, partners' and stakeholders' involvement is recognized as an important joint effort. Partners can be divided into different groups such as science partners (include breeders, engineers, etc.), implementation partners and local stakeholders. Especially for activities focused on women's participation, support from their respective families and communities, especially men, is critical. In such situations, consultation with local stakeholders can benefit the solicitation and vetting of how to implement projects and communicate with societies and target beneficiaries.

Climate finance and gender-responsive budget. Climate financing in developing countries targeting vulnerable communities, including in support of CSA, has seen a major increase in international funding in the past decade (Dinesh *et al.*, 2017; Huang and Wang, 2018; Weiler *et al.*, 2018). Climate projects have been criticized for lack of provision of gender-sensitive climate funding. A report by the Global Greengrants Fund (2015) found that 'only 0.01% of all worldwide grant-dollar support projects address both climate change and women's rights'. Since the 1980s, gender ministries in several countries have led the gender-responsive budget (GRB). Australia, the Philippines and South Africa were three of the earliest initiatives in the 1980s and 1990s. Since the early 2000s, UNIFEM (known as UN Women) started its engagement in GRB work in 20 countries. A gender-responsive budget is 'a budget that works for everyone – women and men, girls and boys – by ensuring gender-equitable distribution of resources and by contributing to equal opportunities for all' (Oxfam, 2018). GRB is not about creating a separate budget for women but aims to ensure allocation of funds that advance gender equality and women's empowerment. Since 2010, there have been international efforts to increase integration of gender into climate projects (Mcleod *et al.*, 2018). Such efforts include the Green Climate Fund (GCF) financial mechanism under the UNFCCC (ibid.). The Green Fund for Development (GFD), in its mandate, states: 'the fund will strive to maximize the impact of its funding for adaptation and mitigation, and seek a balance between the two, while promoting environmental, social, economic and development co-benefits and taking a gender-sensitive approach.'[3]

5. MEAL (Monitoring, Evaluation, Accountability and Learning). 'Frequently, gender gaps are identified in planning, but overlooked in monitoring and evaluation (M&E)' (Huyer *et al.*, 2015). Like other indicators, timely (mid-term, end-of-term, etc.) tracking of progress on gender issues with explicit objectives, strategies, targets and actions is helpful to address gender inequalities and social norms and ensure gender equality and equity. Despite the exclusive efforts to systematize measurement of gender-related goals, there is lack of large-scale and rigorous data available. Most of the data (with local or anecdotal evidence) present male and female participants or users of some technology, leaving an ambiguity about gender equality.

Often, M&E, especially the accountability component, is viewed as donor-driven rather than an essential device and is treated as an 'add-on' rather than a core activity. However, beyond the relationship between project and donors or funding agency, accountability also considers the relationship with beneficiaries and with other local stakeholders, which influences the belief of the beneficiaries. Each component of M&E contributes to learning. Learning from failure as well as success may help for adaptive programming and improving future perspectives. Batliwala and Pittman (2010) have provided a critique of current M&E frameworks and approaches as experienced by women, along with highlighting their strengths and weaknesses in assessing gender equality and women's rights processes and impacts. The authors recommend that MEAL (Monitoring, Evaluation Accountability and Learning) systems should track progress (successes as well as failures) and capture negative impacts, reversals, resistance, reaction and other processes that push back or shift the direction of a positive change trajectory (Fig. 13.2).

6. Action research model. To deal with the complexities in the gender–agriculture–climate nexus, it is of the utmost importance to ensure that the design and

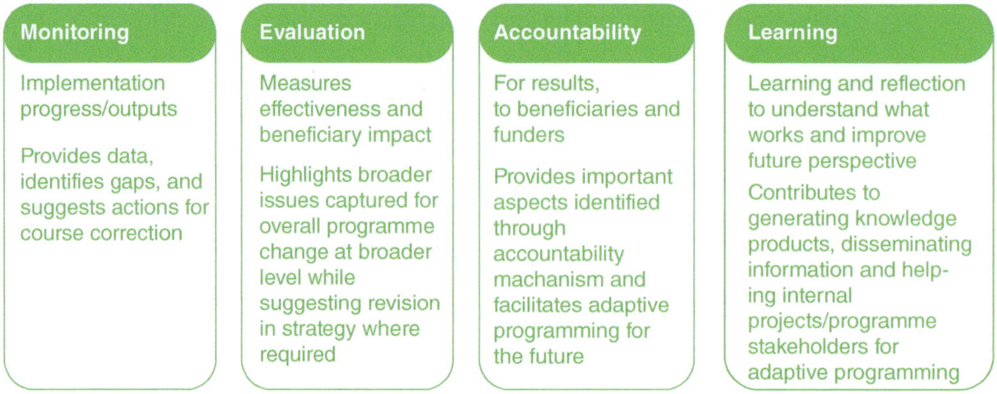

Fig. 13.2. Components of MEAL.

implementation of CRS technologies and practices are actually applicable for end-users. The applicability concern implies that they are tailor-made to the needs and circumstances of the users, affordable and profitable. More specifically, they follow a set of characteristics:

- they are demand-driven and flexible, suited to local contexts and needs;
- they are agreed upon and supported by stakeholders (national to local);
- they provide technical solutions considering the possibility of multiple climate shocks and stages of growth; and
- the intervention programme has an exit strategy, i.e. it builds co-ordination with institutions so that technologies and related input remain available and accessible in the long term. This requires building local mechanisms with self-generating entities for providing CRS-related inputs.

The implementation and adoption of CRS and, in turn, success of CRS projects in the real world with scale and sustainability can be possible with proper planning. Adapting the 'action research spiral' model of Kemmis and McTaggart (2000), we propose a 'four-cycle action research model' (Fig. 13.3). The main advantage of the action research spiral model is the opportunity to understand and analyse the problem in a step-wise manner and in great depth (Kemmis and McTaggart, 2000). The reason for cyclicity and phases in research is that research is context-specific and evolving. At times, the researcher may have to return to previous stages as the research progresses. **Cycle 1: current/existing situation analysis.** Often the failures of plans are blamed on incompatible research context and environmental surroundings. The latter are realized during the final assessment of the plan. So the first step is to choose a focus which determines whether to start a new action (known as proactive action research) or to first collect information relating to possible impacts of the action (known as responsive action research). As the entire process rests upon fundamentals of the research context, it is important to have comprehensive knowledge about gender and climate change issues and mechanisms needed for capacity building. This knowledge can be built with pilot projects, secondary data or information, conducting a small scoping study or by talking to local stakeholders. This means adopting multidisciplinary methods and tools to identify gender issues, knowledge gaps and ways to fill them.

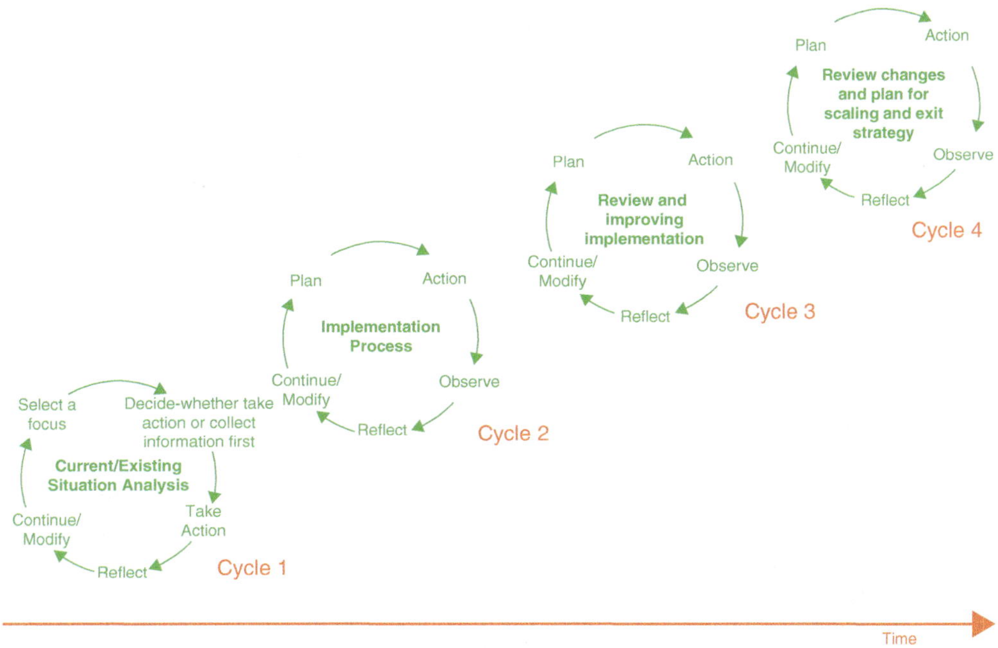

Fig. 13.3. Cycle action research for gender-responsive CRS implementation planning. (From: the author)

Cycle 2: implementation process. The lessons from Cycle 1 will be helpful to map the path for the implementation process. Below are the key points for each of the steps in this cycle:

Plan:

- rationale for mainstreaming gender
- consultative process for gender strategy development
- identify target beneficiary
- gender-strategic or gender-inclusive planning
- gender-accommodative or gender-transformative
- core staffing and institutional capacity (budget)
- cross-connecting teams and themes
- risk and mitigation

Action:

- setting gender-responsive goals
- 'doing' gender
- research questions and activities
- methods and gender-disaggregated data collection
- partnership for catalysing behaviour change

Observe:

- research output and outcome
- impact analysis

Reflect:

- evaluation
- feedback

Cycle 3: review and improve implementation process. Following on from the last step of the previous cycle, this cycle plan will be to revise the implementation process if needed. Often the plans are not executed on the ground for testing. One way to resolve these issues could be to use participatory approaches which may include and train grass roots organizations and local staff.

Cycle 4: review changes and plan for scaling and exit strategy. With this cycle we are aware of how to move forward with the exit strategy of the programme. We are in a better position to understand how to build the mechanisms for the programme to be self-perpetuating.

The four cycles have implicit outcomes: the first cycle's outcome is identifying who needs to be involved; the second cycle's outcome is involving the actors and increasing adoption of strategies; the third cycle's outcome is estimating the impact of strategies; and the last cycle's outcome is sustainable agricultural production and improvement in livelihood.

7. Credibility and validity

Depending on the research method used, appropriate criteria for assessing credibility and validity are applied. Table 13.2 specifies a few validity criteria given by Anderson and Herr (1999).

The above-mentioned validity criteria are appropriate to different circumstances, contexts, nature of the research enquiry (subjective or objective) and preference of generalizations versus extrapolations of research.

Finally, there is the reflexivity in writing the research, especially qualitative research, to be considered. In simple terms, this means how the researcher positions themself in their writing as it influences the reader.

Summary

For engaging women and building their self-confidence to explore new ways of looking at each other as well as new ways of farming that are also climate-enhancing,

Table 13.2. Validity types. (From Anderson and Herr, 1999)

Criteria	Description
Outcome validity	The extent to which actions occur to successfully solve the problem that prompted the research
Process validity	Assessing the adequacy of processes used in different phases of the research
Ecological validity	Involving examination of generalization results from a set of environmental conditions used in present research to another set of conditions
Democratic validity	Refers to the extent to which research is done in collaboration with all parties who have a stake in the problem under investigation.

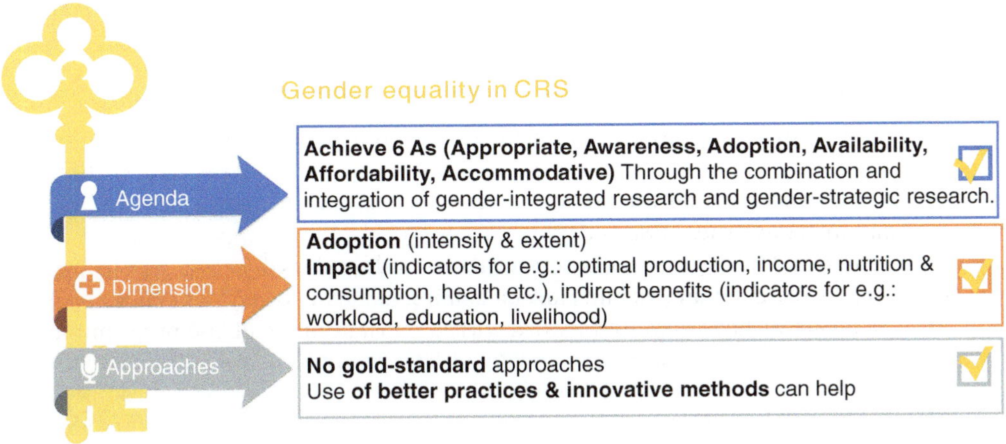

Fig. 13.4. Success key of agenda, dimension and approaches of gender equality in CRS.

gender mainstreaming needs to be revisited and revitalized. This is complex and challenging and requires multiple interventions. Shared vision and values at the local level with beneficiaries and stakeholders can help the implementation process. Gender perspective needs to be integrated into the planning, financing and implementation of adaptation and mitigation efforts by decision makers and implementation agencies and partners. A clear definition and articulation of the beneficiaries' (e.g. farming households, male and female farmers explicitly, beneficiary or user organizations, etc.) agenda with regard to the '6 As' and approaches for gender analysis are needed for gender equality CRS programmes (Fig. 13.4). With the full and equal participation of men and women, a safer and more resilient community development is feasible.

Notes

[1] Transversal means a line that cuts across two or more (usually parallel) lines.

[2] Adapted from http://www.fao.org/climatechange/31734-02e69e384b4ef76bd49477a9650ef4db6.pdf (accessed 27 March 2022)

[3] Source: https://climatefundsupdate.org/the-funds/green-climate-fund/ (accessed 28 March 2022)

References

Alston, M. (2014) Gender mainstreaming and climate change. *Women's Studies International Forum* 47, 287–294.

Aguilar, L. (2016) Foreword. In: Colfer, C.J.P., Basnett, B.S. and Elias, M. (eds) *Gender and Forests: Climate Change, Tenure, Value Changes and Emerging Issues*. Earthscan, London.

Anderson, G.L. and Herr, K. (1999) The new paradigm wars: Is there room for rigorous practitioner knowledge in schools and universities? *Educational Researcher* 28(5), 12–40.

Batliwala, S. and Pittman, A. (2010) *Capturing changes in women's lives: the experiences of Oxfam Canada in applying feminist evaluation principles to monitoring and evaluation practices*. AWID. Available at: https://www.awid.org/publications/capturing-change-womens-realities (accessed 6 June 2022).

Dinesh, D., Aggarwal, P., Khatri-Chhetri, A., Rodríguez, A.M.L., Mungai, C., Sebastian, L. and Zougmore, R.B. (2017) The rise in climate-smart agriculture strategies, policies, partnerships and investments across the globe. *Agriculture for Development* 30, 4–9.

Global Greengrants Fund (2015) *Climate Justice and Women's Rights: A Guide to Supporting Grassroots Women's Action*. Prepared by Ford Foundation Center, International Network of Women's Funds, and Global Greengrants Fund. Available at: https://www.issuelab.org/resources/21050/21050.pdf (accessed 3 June 2020).

Huang, J. and Wang, Y. (2018) Financing sustainable agriculture under climate change with a specific focus on foreign aid. In: Huang, Y. and Pascual, U. (eds) *Aid Effectiveness for Environmental Sustainability*. Palgrave Macmillan, Singapore, pp. 329–367.

Huyer, S., Twyman, J., Koningstein, M., Ashby, J. and Vermeulen, S.J. (2015) *Supporting Women Farmers in a Changing Climate: Five Policy Lessons*. CCAFS Policy Brief 10. CGIAR Research Program on Climate Change, Agriculture and Food Security (CCAFS), Copenhagen.

Kemmis, S. and McTaggart, R. (2000) Participatory action research. In: Denzin, N.K. and Lincoln, Y.S. (eds) *Handbook of Qualitative Research*, 2nd edn. Sage, Thousand Oaks, California.

Mcleod, E., Arora-Jonsson, S., Masuda, Y.J., Bruton-Adams, M., Emaurois, C.O. *et al.* (2018) Raising the voices of Pacific Island women to inform climate adaptation policies. *Marine Policy* 93, 178–185.

Mehar, M., Benzie, J., Mekkawy, W. and McDougall, C. (2019) A systematic literature review of fish trait preferences by end users: implications for fish breeding. *Reviews in Aquaculture* 12, 1273–12956.

Oxfam (2018) *A Guide to Gender Responsive Budgeting*. Oxfam, London. Available at: https://policy-practice.oxfam.org/resources/rough-guide-to-gender-responsive-budgeting-620429/ (accessed 10 June 2022).

UNFCCC (2012) *National Adaptation Plans. Technical Guidelines for the National Adaptation Plan Process*. Least Developed Countries Expert Group, Bonn, Germany.

Weiler, F., Klöck, C. and Dornan, M. (2018) Vulnerability, good governance, or donor interests? The allocation of aid for climate change adaptation. *World Development* 104, 65–77.

Glossary

Disclaimer: *The definitions of terms included here are primarily drawn from IPCC, UN and FAO published sources. We have tried to provide sources for all the terminology.*

Adaptation (p. 17)
Adaptation benefits The avoided damage costs or the accrued benefits following the adoption and implementation of adaptation measures (IPCC AR4, 2007).
Adaptation costs Costs of planning, preparing for, facilitating and implementing adaptation measures, including transition costs (IPCC AR4, 2007).
Adaptive capacity The ability of a system to adjust to climate change (including climate variability and extremes), to moderate potential damages, to take advantage of opportunities, or to cope with the consequences (IPCC AR4, 2007).
Adoption A change in practice or technology used by economic agents or a community. It has a long intellectual history. Adoption is generally measured as a discrete choice and is sometimes associated with a continuous indicator: the extent of adoption (Zilberman *et al.*, 2012).
Agroforestry (p. 132)
Agrosilvicultural system (p. 132)
Apiculture (p. 132)
Aquaforestry (p. 132)
Asset rights These include ownership, use, decision making and documentation over land, housing, material assets, livestock and financial assets (Jacobs *et al.* 2011, p. 31).
Ceteris paribus Latin phrase meaning 'other things held constant or unchanged'.
Climate adaptation (p. 82)
Climate analogue Describes two sites which have similar climatic conditions but are separate geographically and have experienced similar climatic conditions at two different time periods.
Climate change (p. 2, p. 5)
Climate mitigation (p. 80)

Climate variability (p. 6)
Conceptual literature review (p. 99)
Conservation agriculture Has three main principles: minimum soil disturbance; soil protection by vegetation cover; crop rotation and plant association (FAO, 2012).
Consumption smoothing (p. 80)
Coping strategies (p. 80)
Data The actual measurements or factual information that result from the indicator. It could be a number, a fact, an opinion or a perception that points to a specific condition or situation. *Example. Daily temperature, wind speed, monthly rainfall.*

> **Sex-disaggregated data** Data that are collected and analysed separately for men and women.
> *Example: 100 people attend the climate-smart related trainings, out of which 70 are women and the rest are men.*
> **Gender-disaggregated data** In addition to sex-disaggregated data, they reflect other features with differences, commonality, activities, etc., i.e. gender-disaggregated data help us to understand sex-disaggregated data.
> *Example: Out of the 70 women who attended the training and received fertilizer, 35% are from female-headed households who use it on their own field; 50% of the women gave it to their husband; and the remaining 15% sold it to another farming family.*

Disaster Severe alterations in the normal functioning of a community or a society due to hazardous physical events interacting with vulnerable social conditions, leading to widespread adverse human, material, economic or environmental effects that require an immediate emergency response to satisfy critical human needs and that may require external support for recovery (IPCC, 2014). They are commonly categorized as either slow-onset or rapid-onset.

> **Slow-onset disasters** Relates to environmental degradation and can be predicted in advance such as sea-level rise, increasing temperatures, ocean acidification, glacial retreat and related impacts, salinization, land and forest degradation, loss of biodiversity and desertification (Decision 1/CP.16, UNFCCC https://unfccc.int/resource/docs/2010/cop16/eng/07a01.pdf). Droughts, El Niño and La Niña events, and famines are also considered slow-onset disasters.
> **Rapid-onset disasters** Also known as sudden-onset disasters with immediate occurrence and impact. They comprise hydro-metrological hazards such as flooding, windstorms, volcanic eruptions, landslides, earthquakes, tsunamis, hurricanes and forest fires.

Disaster risk reduction (DRR) Denotes both a policy goal and an objective, and the strategic and instrumental measures employed for anticipating future disaster risks; reducing existing exposure, hazard or vulnerability, and improving resilience (IPCC, 2012).
Exposure The presence of people, livelihoods, species or ecosystems, environmental functions, services and resources, infrastructure or economic, social or cultural assets in places and settings that could be adversely affected (IPCC, 2014).

Food security This exists when all people, at all times, have physical, social and economic access to sufficient, safe and nutritious food that meets their dietary needs and food preferences for an active and healthy life. Household food security is the application of this concept at the family level, with individuals within households the focus of concern (EC-FAO, 2008).
Gender (p. 7)
Gender-aware (p. 44)
Gender-expoitative (p. 44)
Gender-transformative (p. 45)
Gender accommodating (p. 7, p. 45)
Gender actors The collective institutions and individuals working on gender in a particular context, including government and academic and civil society actors (NAP Global Network and UNFCCC, 2019, p. 7).
Gender analysis (p. 10)
Gender blind (p. 44)
Gender division of labour (p. 27)
Gender equality (p. 9)
Gender equality continuum (p. 43)
Gender equity (p. 9)
Gender exclusion (p. 7)
Gender expert (p. 51)
Gender gaps (p. 9)
Gender inclusive
Gender integration Refers to strategies applied in activity assessment, design, implementation and evaluation to take gender norms into account and to compensate for gender-based inequalities (Faramand *et al.*, 2017).
Gender mainstreaming (p. 46)
Gender planning (p. 10)
Gender practical needs (p. 10)
Gender-strategic needs (p. 10)
Gender-transformative (p. 45)
Greenhouse gases Those gaseous constituents of the atmosphere, both natural and anthropogenic, that absorb and emit radiation at specific wavelengths within the spectrum of infrared radiation emitted by the earth's surface, the atmosphere and clouds. This property causes the greenhouse effect. Water vapour (H_2O), carbon dioxide (CO_2), nitrous oxide (N_2O), methane (CH_4) and ozone (O_3) are the primary greenhouse gases in the earth's atmosphere. As well as CO_2, N_2O, and CH_4, the Kyoto Protocol deals with the greenhouse gases sulphur hexafluoride (SF_6), hydrofluorocarbons (HFCs) and perfluorocarbons (PFCs) (IPCC AR4, 2007).
Hazard The potential occurrence of a natural or human-induced physical events, trends or physical impacts that may cause loss of life, injury or other health impacts, as well as damage and loss to property, infrastructure, livelihoods, service provision, ecosystems and environmental resources. In this report, the term hazard usually refers to climate-related physical events or trends or their physical impacts (IPCC, 2014).
Impacts (consequences, outcomes) Effects on natural and human systems of extreme weather and climate events and of climate change. Impacts generally

refer to effects on lives, livelihoods, health, ecosystems, economies, societies, cultures, services and infrastructure due to the interaction of climate changes or hazardous climate events occurring within a specific time period and the vulnerability of an exposed society or system. Impacts are also referred to as consequences and outcomes. The impacts of climate change on geophysical systems, including floods, droughts and sea-level rise, are a subset of impacts called physical impacts (IPCC AR4, 2007).

Ex-ante impact Assessment before the event or intervention
Ex-post impact Assessment after the event or intervention

Incentive The material, social or purpose-orientated motivations set for (or held by) individuals, groups and organizations that shape attitudes towards risk and reward (e.g. with respect to performing market functions and/or changing what they do).
Income smoothing (p. 80)
Indicator A particular characteristic or dimension that is observable and measurable and can be used to estimate changes in or progress of a programme. *Example: the number or percentage of women and men participating in climate-smart agriculture-related Farmer Field Schools or farmer-to-farmer extension services (FAO and IFAD, 2015).*

Gender indicators Also referred to as gender-sensitive indicators or gender-responsive indicators, they measure gender-related changes in terms of status and roles of women and men, and changes in gender relations in the household and communities over time (FAO, 2017).

Indigenous people Known as first people in some regions as they are inheritors of a unique culture. Mostly represents tribal groups in this book.
Innovation New or improved behaviour, practice or technique adopted by a market player as a result of programme intervention that confers a benefit to the poor. These can be goods or services and/or new roles that support a different way of working.
Institutions Structures and mechanisms of social, political and economic order and co-operation, either formal or informal, which shape the incentives, capacities and therefore behaviours/practices of market players.
Intervention A defined set of temporary activities through which facilitators seek to effect change in a market system.
Kinship system

Bilateral Refers to the practice of tracing descent and inheritance through both parents (Colfer and Minarchek, 2013, p. 426).
Matrilineal Inheritance and kinship affiliation that typically flows from mother to daughter, with maternal uncles and brothers typically having more authority than in patrilineal systems (Colfer and Minarchek, 2013, p. 426).
Patrilineal Inheritance and kinship affiliation that typically flows from father to son (Colfer and Minarchek, 2013, p. 426).

Morbidity Rate of occurrence of disease or other health disorders within a population, taking account of the age-specific morbidity rates. Morbidity

indicators include chronic disease incidence/prevalence, rates of hospitalization, primary care consultations, disability-days (i.e. days of absence from work), and prevalence of symptoms (IPCC AR4, 2007).
Mortality Rate of occurrence of death within a population; calculation of mortality takes account of age-specific death rates and can thus yield measures of life expectancy and the extent of premature death (IPCC AR4, 2007).
Narrative review (p. 66)
Resilience Amount of change a system can undergo without changing state (IPCC-TAR 2001); the ability of a social or ecological system to absorb disturbances while retaining the same basic structure and ways of functioning; the capacity for self-organization and to adapt naturally to stress and change (IPCC AR4, 2007).
Risk The potential for consequences where something of value is at stake and where the outcome is uncertain, recognizing the diversity of values. Risk is often represented as probability or likelihood of occurrence of hazardous events or trends multiplied by the impacts if these events or trends occur. In this report, the term 'risk' is often used to refer to the potential, when the outcome is uncertain, for adverse consequences on lives, livelihoods, health, ecosystems and species, economic, social and cultural assets, services (including environmental services) and infrastructure (IPCC, 2014).
Scale

> **Macro** Refers to issues of global concern.
> **Meso** Ranges from the national level to a supra-village scale (a district, a particular forest, a landscape).
> **Micro** Ranges from the village through sub-groups within a village to intra-household issues.

Semi-arid regions Regions of moderately low rainfall, which are not highly productive and are usually classified as rangelands. 'Moderately low' is widely accepted as between 100 and 250 mm precipitation per year (IPCC AR4, 2007).
Sensitivity Degree to which a system is affected, either adversely or beneficially, by climate- related stimuli. The effect may be direct (e.g. a change in crop yield in response to a change in the mean, range or variability of temperature) or indirect (e.g. damages caused by an increase in the frequency of coastal flooding due to sea-level rise) (IPCC-TAR, 2001).
Silvopastoral systems (p. 132)
Small and marginal farmers Farmers with ownership of less than 1 hectare of land.
Stakeholder A person or an organization that has a legitimate interest in a project or entity, or would be affected by a particular action or policy (IPCC AR4, 2007).
Stocktaking (p. 176)
Sustainable agriculture A sustainable agriculture is one that, over the long term, enhances the environmental quality and the resource base on which agriculture depends; provides for basic human food and fiber needs; is economically viable; and enhances the quality of life for farmers and society as a whole (Francis and Youngberg, 1990).
Sustainable development Development that meets the cultural, social, political and economic needs of the present generation without compromising the ability of future generations to meet their own needs (IPCC AR4, 2007).

Systematic literature review Systematic literature review is an exhaustive and robust review and synthesis of research evidence which seeks to systematically search for, appraise and synthesise research evidence, often adhering to guidelines on the conduct of a review (Grant and Booth, 2009).

Time poverty Can be understood as the fact that some individuals do not have enough time for rest and leisure after taking into account the time spent working, whether in the labour market, for domestic work, or for other activities such as fetching water and wood (Bardasi and Wodon, 2006).

Undernourishment Exists when caloric intake is below the minimum dietary energy requirement (MDER) (FAO, 2009). The MDER is the amount of energy needed for light activity and a minimum acceptable weight for attained height, and it varies from country to country and from year to year depending on the gender and age structure of the population. Throughout the report by FAO (2009), the words 'hunger' and 'undernourishment' are used interchangeably.

United Nations Framework Convention on Climate Change (UNFCCC) Adopted on 9 May 1992, in New York and signed at the 1992 Earth Summit in Rio de Janeiro by more than 150 countries. Its ultimate objective is the 'stabilisation of greenhouse gas concentrations in the atmosphere at a level that would prevent dangerous anthropogenic interference with the climate system' (IPCC AR4, 2007).

Vector-borne diseases Diseases that are transmitted between hosts by a vector organism (such as a mosquito or tick); e.g. malaria, dengue fever and leishmaniasis (IPCC AR4, 2007).

Vulnerability The degree to which a system is susceptible to, or unable to cope with, adverse effects of climate change, including climate variability and extremes. Vulnerability is a function of the character, magnitude and rate of climate variation to which a system is exposed, its sensitivity, and its adaptive capacity (IPCC AR4, 2007).

References

Bardasi, E. and Wodon, Q. (2006) *Measuring time poverty and analyzing its determinants: concepts and application to Guinea*. MPRA Paper 11082, University Library of Munich, Germany.

Colfer, C.J.P. and Minarchek, R.D. (2013) Introducing 'the gender box': a framework for analysing gender roles in forest management. *International Forestry Review* 15(4), 411–426.

EC-FAO (2008) *An Introduction to the Basic Concepts of Food Security*. EC - FAO Food Security Programme, Food and Agriculture Organization of the United Nations, Rome, Italy.

FAO (2008a) Investing in sustainable crop intensification: the case for improving soil health. *Integrated Crop Management* 6.

FAO (2008b) *Deriving Food Security Information from National Household Budget Surveys: Experiences, Achievements, Challenges*. FAO, Rome.

FAO (2009) *The State of Food Insecurity in the World: Economic Crises-impacts and Lessons Learned*. Food and Agriculture Organization of the United Nations, Rome, Italy.

FAO (2012) What is CA? Principles of CA, Benefits of CA. Available at: http://www.fao.org/ag/ca/index.html (accessed July 2017).

FAO (2017) Conservation agriculture. Available at: http://www.fao.org/ag/ca/index.html (accessed December 2018).

FAO and IFAD (2015) Gender in climate-smart agriculture. Module 18 for *Gender in Agriculture Sourcebook. Agriculture Global Practice*. World Bank, Washington, DC.

Faramand T.H., Ivankovich, M. and Holtemeyer, J. (2017) *A Guide to Integrating Gender in Improvement*. USAID Applying Science to Strengthen and Improve Systems (ASSIST) Project. University Research Co., LLC, Chevy Chase, Maryland. Available at: https://www.urc-chs.com/wp-content/uploads/urc-assist-integrating-gender-guide.pdf (accessed 10 March 2022).

Francis, C.A. and Youngberg, G. (1990) Sustainable Agriculture - an overview. In: Francis, C.A., Flora, C.B. and King, L.D. (eds) *Sustainable Agriculture in Temperate Zones*. John Wiley and Sons, New York, pp. 1–23.

Grant, M.J. and Booth, A. (2009) A typology of reviews: an analysis of 14 review types and associated methodologies. *Health Information & Libraries Journal* 26, 91–108.

IPCC (2012) *Managing the Risks of Extreme Events and Disasters to Advance Climate Change Adaptation*. A Special Report of Working Groups I and II of the Intergovernmental Panel on Climate Change [Field, C.B., Barros, V., Stocker, T.F., Qin, D., Dokken, D.J. *et al.* (eds.)]. Cambridge University Press, Cambridge, UK and New York.

IPCC (2014) Summary for policymakers. In: *Climate Change 2014: Impacts, Adaptation, and Vulnerability.* Part A: Global and Sectoral Aspects. Contribution of Working Group II to the Fifth Assessment Report of the Intergovernmental Panel on Climate Change (C.B. Field *et al.* (eds)). Cambridge University Press, Cambridge, UK and New York.

IPCC AR4 (2007) *Climate Change 2007: Synthesis Report*. Contribution of Working Groups I, II and III to the Fourth Assessment Report of the Intergovernmental Panel on Climate Change [Core Writing Team: Pachauri, R.K and Reisinger, A.]. IPCC, Geneva, *Switzerland.*

IPCC TAR (2001) *Climate Change 2001: Impacts, Adaptation and Vulnerability*. IPCC Third Assessment Report. Cambridge University Press, Cambridge, UK.

Jacobs, K., Namy, S., Kes, A., Bob, U. and Moodley, V. (2011) *Gender Differences in Asset Rights in KwaZulu-Natal, South Africa.* International Center for Research on Women, Washington, DC.

NAP Global Network and UNFCCC (2019) Toolkit for a gender-responsive process to formulate and implement National Adaptation Plans (NAPs). (Dazé, A. Church, C. (lead authors)). International Institute for Sustainable Development, Winnipeg, Canada. Available at: from www.napglobalnetwork.org (accessed 10 March 2022).

Zilberman, D., Zhao, J. and Heiman, A. (2012) Adoption versus adaptation, with emphasis on climate change. *Annual Review of Resource Economics* 4(1), 27–53.

Reference Guides and Websites

Tools and Guides

Bonilla-Findji, O., Eitzinger, A.N. and Andrieu, N. (2021) *Implementation Manual. CCAFS Climate-smart Monitoring Framework: Tackling Uptake of CSA Options and Perceived Outcomes at Household and Farm Level.* CGIAR Research Program on Climate Change, Agriculture and Food Security (CCAFS). Wageningen, The Netherlands.
This manual develops an integrated CSA monitoring framework, which provides a set of core and extended indicators linked to the research question at household level.

Britton, B., Lipson, B. and Cronin, O. (2015) *Compendium of Qualitative Methods for Monitoring and Evaluating Organizational Development.* WWF, Gland, Switzerland.
This document examines six methods and four tools that are particularly suited to monitoring and evaluating (M&E) organizational development (OD). The methods include: stories of change, most significant change, outcome mapping, appreciative inquiry, Sensemaker®, organizational assessment.

FAO (2011) *Training Guide Gender and Climate Change Research in Agriculture and Food Security for Rural Development*. Food and Agriculture Organization of the United Nations, Rome.
This training guide provides a clear understanding of the concepts related to gender and climate-smart agriculture. It describes participatory methods for conducting gender-sensitive research on the impacts of climate change.

FAO (2013) *Climate-smart Agriculture Sourcebook*. Food and Agriculture Organization of the United Nations, Rome.
This sourcebook is a reference tool for planners, practitioners and policy makers working in agriculture, forestry and fisheries at national and subnational levels.

FAO (2017) The role of gender in climate-smart agriculture. Module C-6 in the *Climate Smart Agriculture Sourcebook*. Food and Agriculture Organization of the United Nations, Rome.
The module synthesizes recent research evidence and experiences with climate-smart agriculture to provide guidance to a wide range of stakeholders on opportunities for future gender-responsive climate-smart agricultural investments and interventions.

FAO (2018) *Climate-smart Agriculture Training Manual: A Reference Manual for Agricultural Extension Agents.* Food and Agriculture Organization of the United Nations, Rome.
This manual is intended for agricultural extension agents and others interested in developing training courses on climate-smart agriculture.

FAO (2018) Guidance note on gender-sensitive vulnerability assessments in agriculture. Food and Agriculture Organization of the United Nations, Rome.
This guidance note provides an overview of available sources and methodologies to collect and analyse sex-disaggregated data.

FAO (2019) *Climate-smart Agriculture and the Sustainable Development Goals: Mapping Interlinkages, Synergies and Trade-offs and Guidelines for Integrated Implementation.* Food and Agriculture Organization of the United Nations, Rome.
This publication presents an assessment and mapping of CSA-SDG interlinkages. These provide entry points for targeted CSA planning, to enhance synergies and reduce potential trade-offs between CSA objectives and SDGs. The publication also provides guidelines for integrating the CSA implementation steps with the 2030 Agenda.

FAO and CARE (2019) Good practices for integrating gender equality and women's empowerment in climate-smart agriculture programmes.
This paper documents a set of climate-smart agriculture practices, integrating with gender equality and women's empowerment (GEWE) from agriculture, agroforestry and livestock and fisheries sector.

FAO and World Bank (2017) *How to Integrate Gender Issues in Climate-smart Agriculture Projects.*
This training manual provides ways of integrating gender throughout the project cycle with a glossary of key terms and concepts of CSA and gender, checklists for gender integration, approaches and tools for conducting gender analysis and lessons from previous works.

NAP Global Network and UNFCCC (2019) *Toolkit for a Gender-responsive Process to Formulate and Implement National Adaptation Plans (NAPs).* International Institute for Sustainable Development, Winnipeg, Canada.
This toolkit is designed to support country efforts to pursue a gender-responsive NAP process. It will be useful for government actors co-ordinating the NAP process, as well as for stakeholders and development partners supporting adaptation planning and implementation.

Nelson, S. and Hill, C. (2019) *Gender in Adaptation Planning for the Agriculture Sectors: A Guide for Trainers*. Food and Agriculture Organization of the United Nations, Rome.
This guide provides a complete set of materials for use in training sessions on mainstreaming gender in adaptation planning in the agricultural sectors.

Ngara, T. (2017) *Climate-smart Agriculture Manual for Zimbabwe*. Climate Technology Centre and Network, Copenhagen.
The manual presents the entry points for introducing CSA interventions under three thematic areas: enabling environments, system approaches, practices.

Oxfam (2018) *A Guide to Gender-responsive Budgeting*.
This guide provides a typical budget cycle diagram which frames the different actions that different actors (i.e. government, civil society or international NGOs) can use for gender-responsive budgeting.

SERVIR-Mekong (2015) Gender and GIS guidance notes.
This guide is targeted at geographic information systems (GIS) application developers and potential application users, and intends to provide step-by-step guidance on integrating gender concerns in the production of GIS applications for environmental/natural resource management, disaster risk management and building resilience to climate change.

UNDP (2010) *Gender, Climate Change and Community-Based Adaptation*. United Nations Development Program, New York.
This guide provides lessons learned from the UNDP-Global Environment Facility (GEF) Community-based Adaptation Programme and from the GEF Small Grants Programme in over 122 countries.

World Bank, FAO and IFAD (2015) Gender in climate-smart agriculture: Module 18 of the *Gender in Agriculture Sourcebook*.
This module includes tested good practices, innovative approaches and technologies for gender mainstreaming in CSA.

Zusman, E., Young, L., Rojas, A. and Adams, L. (2016) Mainstreaming gender into climate mitigation activities: guidelines for policy makers and proposal developers. Asian Development Bank. Mandaluyong City, the Philippines.
This document offers guidance to policymakers, proposal developers and other stakeholders to mainstream gender into mitigation actions and funding proposals.

Websites

Climate-L.org: http://climate-l.org/
A knowledge base of United Nations and intergovernmental activities addressing global climate change policy.

Climate Smart Agriculture: https://csa.guide/csa/what-is-climate-smart-agriculture
This portal was developed by the by the CGIAR Research Program on Climate Change, Agriculture and Food Security (CCAFS) for the World Bank in collaboration with a range of other partners and institutions. This contains information from basics to practical practices.

FAO Gender: https://www.fao.org/gender/resources/en/
This page compiles all information for FAO projects focusing gender.

Gender Climate Tracker App: https://wedo.org/gender-climate-tracker-app/
By compiling policies, mandates, research, decisions and actions related to gender and climate change, the app aims to highlight key gender entry points to facilitate implementation.

GenderCC network – women for climate justice: www.gendercc.net
A global network of organizations, experts and gender activists from all world regions working for gender and climate justice.

Gender and Disaster Network: www.gdnonline.org
The Gender and Disaster Network (GDN) is an educational and advocacy project initiated by women and men interested in gender relations in disaster contexts. The GDN seeks to promote gender equity on access to and safe management of water sources. This web portal contains a wide range of information and tools to work on the theme, as well as case studies, projects and policies.

Gender and social inclusion: https://ccafs.cgiar.org/research/gender-and-social-inclusion
This portal is a resource hub for all work done under CCAFS projects. The work includes information about projects, publications, gender and social inclusion tools, outcomes, events and news.

Global Gender and Climate Alliance (GCCA): https://www.gcca.eu
The Global Gender and Climate Alliance (GCCA) is a European Union flagship initiative and was launched in 2007 by the UNFCCC COP13 meeting by UNDP, UNEP, IUCN and WEDO to propel gender into the climate agenda.

International network on gender and sustainable energy: www.energia.org
This is an international network, established in 1996 by a group of energy experts from developing countries. This has information on gender and energy, including academic articles, case studies and strategies for action.

UN WOMEN Climate Change and Environment: https://www.unwomen.org/en/how-we-work/intergovernmental-support/climate-change-and-the-environment
UN Women works for women's active participation, as well as for attention to gender issues and the inclusion of gender-specific mandates across negotiated outcomes and agreements in the context of the three United Nations Conventions on climate change, desertification, and biodiversity, and other multilateral environmental instruments. UN Women prepares research and technical papers to inform deliberations of Parties, convenes meetings with governments and other stakeholders, and supports the participation of gender equality advocates in key gatherings. The portal has detailed information for all this work.

UNFCCC documents related to the Gender and Climate Change agenda item: https://unfccc.int/topics/gender/resources/documentation-on-gender-and-climate-change
This web page contains all the documents (i.e. COP decisions, reports, synthesis reports, workshop reports, technical paper, informal/formal note, presentations) related to the Gender and Climate Change agenda item.

Women's Environment & Development Organization (WEDO): http://www.wedo.org/
This platform connects and facilitates women's organizations, governments and international organizations, including the United Nations, to ensure that women's human rights are at the heart of global and national policies.

Index

Note: Page numbers in **bold** type refer to **figures**; Page numbers in *italic* type refer to *tables* or *boxes*